The Open University

MU120
Open Mathematics

Unit 16

# Rainbow's end

MU120 course units were produced by the following team:

Gaynor Arrowsmith (Course Manager)
Mike Crampin (Author)
Margaret Crowe (Course Manager)
Fergus Daly (Academic Editor)
Judith Daniels (Reader)
Chris Dillon (Author)
Judy Ekins (Chair and Author)
John Fauvel (Academic Editor)
Barrie Galpin (Author and Academic Editor)
Alan Graham (Author and Academic Editor)
Linda Hodgkinson (Author)
Gillian Iossif (Author)
Joyce Johnson (Reader)
Eric Love (Academic Editor)
Kevin McConway (Author)
David Pimm (Author and Academic Editor)
Karen Rex (Author)

Other contributions to the text were made by a number of Open University staff and students and others acting as consultants, developmental testers, critical readers and writers of draft material. The course team are extremely grateful for their time and effort.

The course units were put into production by the following:

Course Materials Production Unit (Faculty of Mathematics and Computing)

Martin Brazier (Graphic Designer)
Hannah Brunt (Graphic Designer)
Alison Cadle (TEXOpS Manager)
Jenny Chalmers (Publishing Editor)
Sue Dobson (Graphic Artist)
Roger Lowry (Publishing Editor)
Diane Mole (Graphic Designer)
Kate Richenburg (Publishing Editor)
John A.Taylor (Graphic Artist)
Howie Twiner (Graphic Artist)
Nazlin Vohra (Graphic Designer)
Steve Rycroft (Publishing Editor)

This publication forms part of an Open University course. Details of this and other Open University courses can be obtained from the Student Registration and Enquiry Service, The Open University, PO Box 197, Milton Keynes MK7 6BJ, United Kingdom: tel. +44 (0)845 300 6090, email general-enquiries@open.ac.uk

Alternatively, you may visit the Open University website at http://www.open.ac.uk where you can learn more about the wide range of courses and packs offered at all levels by The Open University.

To purchase a selection of Open University course materials visit http://www.ouw.co.uk, or contact Open University Worldwide, Walton Hall, Milton Keynes MK7 6AA, United Kingdom, for a brochure: tel. +44 (0)1908 858793, fax +44 (0)1908 858787, email ouw-customer-services@open.ac.uk

The Open University, Walton Hall, Milton Keynes, MK7 6AA.

First published 1996. Second edition 2001. Third edition 2004. New edition 2008.



Edited, designed and typeset by The Open University, using the Open University TEX System.

Printed and bound in the United Kingdom by The Charlesworth Group, Wakefield.

ISBN 978 0 7492 2873 6

4.1

# Contents

# Study guide

This unit is designed to help you to revise and consolidate your work on the whole course. You should study Section 1 first. It will assist you in planning your work on the rest of the unit and on the final consolidation assignments (TMA04 and CMA45). Sections 2 and 3 may then be studied in whichever order you prefer, depending upon when you find it convenient to watch the video associated with Section 2.

During your work on Section 1, you will need to look at the details of the consolidation assignments so you can appreciate what is involved, and then decide what you need to do for CMA45, make a preliminary choice from the options in TMA04 and plan accordingly. There is an activity sheet you may find helpful for your plan.

The television programme, *The Rainbow*, is very relevant to this unit. It is best viewed *prior* to working on Section 2, which looks at mathematical models of the rainbow. Section 2 revises much of the mathematics from the course, as well as the modelling and calculator techniques. Several of the activities should be relevant to TMA04. You should watch the separate video as you work through this section.

Section 3 consolidates your calculator work on the course, and should be particularly useful if you choose the programming question in TMA04. It directs you to Chapter 16 of the *Calculator Book*, which reviews most of the programming commands on your calculator and suggests ways of building up a library of programs to suit your mathematical needs.

Section 4 includes an audio band revisiting 'Mathematical musings' from *Unit 1*. This time, listen with the insights you have gained from your mathematical journey through MU120, recalling aspects from the whole course and your progress through it. This should help you with Part 2 of TMA04.

Section 5 is very closely linked to TMA04. You will need to look through your assignments, completed activities and the other work on the course in order to select evidence of your mathematical learning and examples of your progress in different aspects. While studying this section, you will need to have available as much of the course material and your work on it as possible.

You will need the assignment booklet which includes both the final assignments while studying Sections 1 and 5, and the *Calculator Book* for Section 3. As this is a consolidation unit, your Handbook notes from the whole course will be useful throughout.

Working through this unit should give you a good start on the final assignments. You should complete them over the next couple of weeks and submit them in plenty of time to reach the Open University by the cut-off dates.

# Introduction

With this final unit, you will complete your mathematical journey through MU120. The main aims of the unit are to provide opportunities for you to use and consolidate your mathematical learning and to reflect on your progress over the course. It is hoped that you will find this final part of the journey interesting and rewarding as you explore the rainbow and look for the 'crock of gold' at its end.

In any course of study, it is easy to get caught up with the details rather than be concerned with the big picture. However, from time to time it is important to stand back and review what you have done and learned overall. You have been encouraged to do this at the end of each block, and now the end of the course is another appropriate time to consolidate your learning.

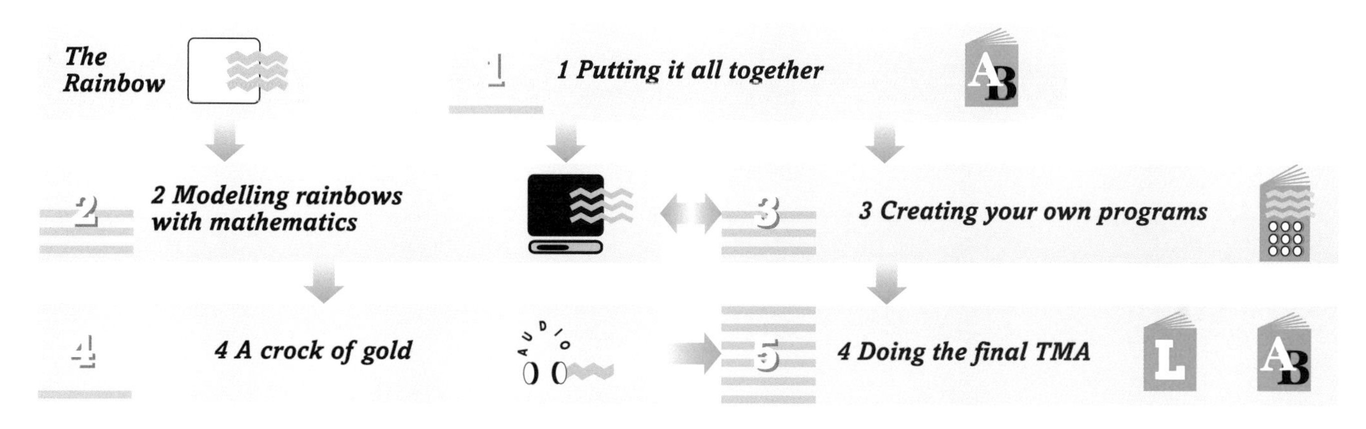


Summary of sections and other course components needed for *Unit 16*.

# 1 Putting it all together

*Aims* The main aims of this section are to help you to appreciate what you need to do for the consolidation assignments and to assist you in planning your work for them and the rest of this unit. ◇

## 1.1 Why put it all together?

MU120 does not have an examination, but it does have final consolidation assignments, which are intended to help you to pull the different threads of the course together. This consolidation process will involve reviewing and revising what you have learned over the course. The process should not just be a memorizing task; rather, it should be a more constructive activity that rounds off your work on MU120: reconstructing ideas, techniques and skills in a more explicit, usable, coherent and enduring form. Revision should be an active process, not mechanical 'scanning through pages, hoping that something will stick', but rather engaging and thought-provoking.

Both the final assignments (TMA04 and CMA45) cover the whole course. There is a lot to do before the cut-off dates. Appreciating what you have to do will enable you to draw up a realistic plan to get everything done in time. Get out the relevant assignment booklet and skim through both assignments to see what is involved.

Remember, you will be putting together your work not only for your tutor and the University, but also for yourself, so you can see what you have achieved in your studies. You will probably be surprised and pleased by how much you have learned.

## 1.2 Reviewing the course for the assignments

The final consolidation assignments may seem slightly daunting, but the work can be broken down into steps.

### Reviewing mathematical themes

In order to tackle the CMA questions, you will most likely need to revise some of the ideas and techniques from earlier in the course. For the TMA, you will need to take a broader view and consider common themes that run through the course—themes such as numerical relationships, or programming the course calculator.

### *Researching topics*

For each of the topics that you choose for TMA04, you will need to find material in the course that is relevant. Bear in mind that a topic may crop up in several different units and in the *Calculator Book*. The indexes and the unit learning outcomes may help you to decide where to focus your research.

### *Selecting evidence of learning*

For many of the questions in TMA04 you will need to use examples of your work as evidence of your achievements: for example, a successfully completed relevant activity (but not just an answer copied from the back of the book). Your CMAs and TMAs are rich sources of evidence. Other possible sources are: Resource Book exercises, tutorial activities and your own notes. You will need to select the evidence that shows clearly that you understand a particular concept or that you can use a specific technique. Quality, not quantity, is important.

### *Presenting work to show achievements*

When you have selected the evidence of your achievements, you will need to organize and present it. Be sure to label the examples of your work clearly, identifying the source and the particular concept or technique.

### *Reflecting on progress*

Part of TMA04 asks you to reflect upon your progress during your study of the course. You will need to choose one aspect of your study where you are able to:

◇ identify what you have learned;

◇ describe your progress in this aspect during the course;

◇ consider factors that have influenced your progress.

A number of activities throughout MU120 have prompted you to monitor your work, think about the skills you have learned, review your progress, record ideas and use what you have learned. So you may have relevant completed activity sheets and notes to help you, as well as your TMAs.

## *1.3 Planning your consolidation assessment work*

Look at TMA04 and CMA45 and think about how you will tackle them.

The CMA consists of short questions covering the whole course. Obviously you do not have time to re-read the course from start to finish in order to answer these questions. You will need to decide upon a strategy.

## Activity 1 Your CMA strategy

Read through the CMA and divide the questions into three categories: those you might be able to do without referring to your notes or to the course materials; those you could probably answer by referring back to these items; and those where you feel you would need to look back over the course in more detail before you could answer.

Write down a list of what you need to do in order to complete CMA45.

Next turn to TMA04 and consider the choices that you will need to make. In Part 1, you will have to choose from questions relating to themes that have run through the course. Some questions require you to select pieces of your own work as evidence of your mathematical achievement and to write an explanation of the concepts and/or techniques involved. However, some questions (like the calculator question) are a bit different. You will also have to choose an aspect of your study for the reflective question in Part 2.

## Activity 2 Your strategy for TMA04

Read TMA04 carefully.

(a) Look at Part 1 of the TMA. Think about which questions you might be able to tackle. Write notes on what you would need to do for each.

Try to make a provisional choice now, for the purpose of planning your work. You can change your mind later if you wish.

(b) Go on to Part 2 of the TMA. Think about which aspect you might choose. Write notes on what you would need to do for each aspect. Make a provisional choice of aspect.

## Activity 3 Planning your study of the rest of this unit

Look back at the diagram in the Study guide to this unit and think about how much time you might spend on each section. The detail with which you need to study some sections may depend upon your choice of TMA04 questions. For example, if your choice includes the calculator question, then you may need to spend more time on Section 3 of this unit.

Jot down how much time you think you will need to study each of the sections in this unit.

## *Activity 4* Planning your final assignments

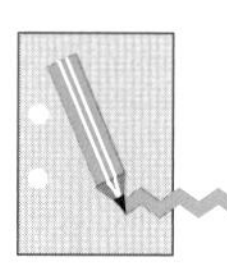

There is an activity sheet that you may find helpful in developing a plan for tackling the final assignments.

From your notes on Activities 1–3, write a list of all the things that you will need to do to complete the assignments and your study of this unit, preferably in the order in which you will do them. Remember to include:

- your study of the other sections of this unit;
- doing CMA45 and, if necessary, revising some units for particular questions;
- researching the topics for TMA04 questions;
- selecting evidence from your work on appropriate topics;
- photocopying the selected examples of your work;
- writing the required explanations on the topics and/or the calculator program;
- doing or redoing calculations/calculator work;
- background research on your chosen aspect for the reflective question;
- writing the reflective account;
- organizing and presenting your work;
- posting TMA04 and CMA45, including obtaining proof of posting.

Against each item in your list, write down how long you think it will take.

Next consider your other commitments and the time you have available to spend on MU120.

Then work out a timetable for doing everything. Try to include some extra time as a contingency against the unexpected. Remember, some things may take you longer than you anticipate, so decide which items in your list are of high priority and those which could be abandoned if you run into unforeseen problems. Indicate the high priority items on your overall plan.

---

You may feel that you do not have enough time to do everything or that there are other uncertainties—think about them and, if necessary, modify the plan. Although your plan may not work out very well in practice, it should give you an idea of what you need to do and the timescale involved.

Try to discuss your plan with your tutor and, perhaps, with other students. Explain your choices and discuss how realistic your ideas are. For instance, you may be attracted to the calculator question but, if you have problems with the program, is there a fall-back position?

Although you want to do well on CMA45, it has a lower weighting than TMA04. So if you cannot answer all the CMA questions and do not have time to revise the relevant units, you could still pass the course, provided you have a reasonable mark on TMA04. However, do submit the CMA, answering as many questions as possible, because even a few marks from it are better than none.

In TMA04, you may feel less confident about one of your chosen questions and think that you need to do a lot of work on it. But do not let the other questions suffer. Remember, you may be able to pass even if you have to omit a question. Mark off items on your plan as you complete them and modify the plan if necessary.

You need to reach a minimum standard both in the course assessment as a whole and in TMA04. You can check your assessment score so far in the activity below and revise the concept of a weighted mean at the same time!

## Activity 5 Calculating your current assessment score

Your assessment score is the weighted mean of your assignment scores. Look in the *Course Guide* for the weightings of the assignments.

Fill in the table below to calculate your own score. At the moment, put in a zero for any score that is not available.

CMA41 is zero weighted so there is no need to include it.

| Assignment | Weighting $w$ | Score $x$ | $wx$ |
|---|---|---|---|
| CMA42 | | | |
| CMA43 | | | |
| CMA44 | | | |
| CMA45 | | 0 | 0 |
| TMA01 | | | |
| TMA02 | | | |
| TMA03 | | | |
| TMA04 | | 0 | 0 |
| Total | | – | |

Calculate your assessment score from $\dfrac{\sum wx}{\sum w}$.

If you have missed an assignment or done badly in some, you might like to consult your tutor about the substitution mechanism. However, remember that there is no substitution for TMA04 or CMA45.

## Outcomes

Now you have completed your study of this section, you should:

- appreciate what is involved in the final consolidation assignments;
- have planned your study of this unit and your work on these;
- have revised the concept of a weighted mean by calculating your assessment score.

# 2 Modelling rainbows with mathematics

*Aims* The main aim of this section is to consider the rainbow in mathematical terms. Much of the mathematics introduced during the course is used here, so the section also serves as revision. ◇

The theme of using mathematics to look at the world runs through the course. Mathematical modelling provides a framework for thinking about problems, and mathematics provides tools for solving these problems. However, phenomena in the physical world do not come with problems ready-posed: it is people who raise questions, sometimes for practical purposes or sometimes just out of curiosity. Bear in mind that looking at a phenomenon mathematically includes trying to decide which sorts of questions can be tackled using the available mathematics.

Applying mathematics to the physical world means treating objects in the physical world as if they were mathematical objects under certain circumstances. Think back to the *Unit 1* reading where John Clearwater saw cabbages as spheres, mountains as triangles, and rivers as flowing in straight lines. A mathematical model stresses the similarities and ignores the differences between physical phenomena and abstract mathematical objects, but the differences need to be kept in mind when interpreting the model.

Recall the mathematical modelling cycle introduced in *Unit 10* and shown in Figure 1 below: for the *purpose* of solving a problem, assumptions are made in order to *create a model*; this mathematical model can then be solved by *doing mathematics*; next the *results are interpreted* in light of the assumptions; finally the *model is evaluated* to see if it is good enough. Often the assumptions are then modified to give an improved model, and the modelling cycle is repeated. Historically, this has been the case with modelling rainbows, as you will see when you work through this section.

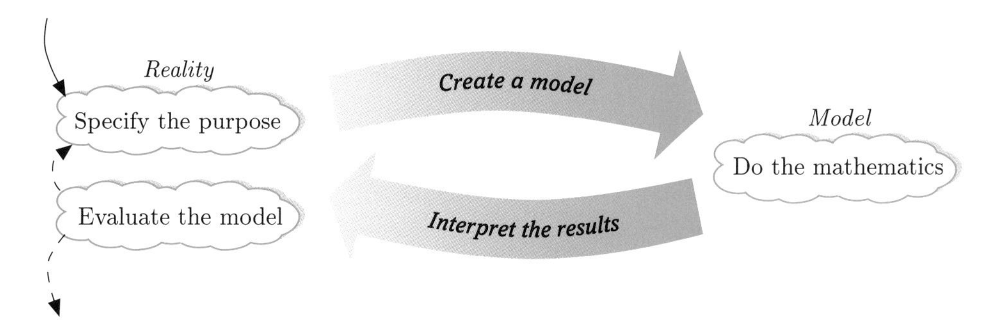


*Figure 1* The mathematical modelling cycle.

To gain an understanding of the rainbow phenomenon, some investigation of data is required alongside the modelling cycle. In *Unit 5*, a four-stage investigation cycle was suggested as a structure to guide such studies. As Figure 2 shows, the cycle consists of *posing the question*, *collecting the data*, *analysing the data*, and *interpreting the results*. *Units 2* and *3* implicitly used this cycle in investigating whether people in Britain are becoming better off; *Unit 4* investigated clusters of diseases in this way, and *Unit 5* considered bird populations on Skomer. In the present unit, modelling rainbows will involve the statistical investigation of data about the behaviour of light as it travels from air into water (raindrops) and then out again.

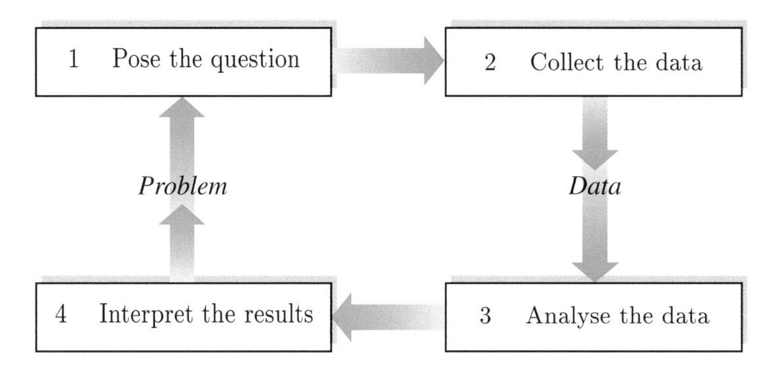


*Figure 2* The statistical investigation cycle.

In modelling the rainbow, many concepts and techniques from earlier in the course will be used, so you may want to check up on some things from previous units, perhaps by looking at your blue Handbook activity sheets or at other notes. In particular, the related concepts of ratio and index will be relevant. Recall how the data in Block A on prices and earnings were analysed using price and earnings ratios and indices. So comparisons are made on *relative differences* (ratios) rather than on *absolute differences* (based on subtraction). This makes the concept more widely applicable. In this section you will come across another use of a relative comparison: an index to describe how light rays travel from air into water and vice versa—a process that is crucial to the formation of rainbows.

## 2.1 Starting to model the rainbow

You probably know quite a lot about rainbows already. Under certain conditions you see bands of coloured light extending in an arc across the sky. Rainbows only appear when the Sun is shining and there is rain or some other type of water droplets in the air. In order to see a rainbow, the Sun must be behind you and the rain in front. Rainbows always have a curved shape, colours in the same order. Sometimes you can see two bows, the outer one being less distinct. A crock of gold is supposedly buried where the rainbow ends—but the end of a rainbow is virtually impossible to find because, as you approach it, it appears to move away.

How can mathematics help your understanding of this phenomenon? Direct experience of the rainbow is one of light, colour and, perhaps, wonder; mathematical explanations make use of angles, ratios, lines, circles and idealized geometrical shapes—they use mathematical models.

Let us start by considering the first stage of the mathematical modelling cycle: *specifying the purpose* of the model. Put simply, the purpose here is to explain the existence of the rainbow. More precisely, the model needs to answer at least some of the many questions about the rainbow, such as:

- ◇ Why does a rainbow appear at all?
- ◇ Why is it the shape it is?
- ◇ Why is it in the position it is?
- ◇ Why is it made up of bands of colours, always in the same order?
- ◇ Why is there sometimes a second rainbow?

Moving on to the second stage of the modelling cycle, begin to think about *creating a model* of the rainbow. Rainbows appear when sunlight falls on raindrops or on a spray of water, such as that from a waterfall or fountain. So any model of the rainbow needs to take into account the interaction of light and drops of water. Therefore consider how light and raindrops might be represented in a mathematical model. Raindrops are roughly spherical, hence a reasonable modelling assumption is that they are exactly spherical. Rays of light are often represented by lines, so drawing diagrams of rays of light passing through a sphere is a good place to start. Since paper is two-dimensional, it is easier to simplify and start by dealing with a two-dimensional circular raindrop, and later to extend the model to three dimensions. As the model will be concerned with the position of the rainbow, angles will be needed. This means that the geometry of lines, angles and circles, as covered in *Unit 14*, will be needed.

Similar phenomena sometimes occur when light passes through other transparent objects, but for the moment just consider water.

This is not the only model of light. Sometimes light is modelled as travelling in waves, analogous to sound waves, and sometimes as a stream of particles or packets of energy, called quanta.

Before you can proceed further with the modelling cycle and look at the development of a specific model of the rainbow, you need to know something about what happens to light rays when they pass from air into water and back out again. This fundamental problem occupied scientists for several centuries and prompted, what were in effect, statistical investigations by Aristotle, Kepler, Snell, Descartes and Newton, among others. So it is worth taking some time now to think about the sort of statistical investigation involved.

In the context of a study of how light rays behave when they pass from air into water, the statistical investigation cycle suggests the following stages:

*Posing the question*

Historically, questions have focused on finding a relationship between the directions of light rays before and after they encounter a boundary between air and water. However, before light rays that pass through the curved surface of a spherical raindrop are investigated, an easier case is usually considered: that of light rays encountering a flat water surface.

Thus the first question in the statistical investigation should be: what is the relationship between the directions of light rays before and after meeting a flat air–water boundary?

*Collecting the data*

A considerable amount of experimental data is already available concerning the directions (angles) of rays of light before and after they meet air–water and water–air boundaries. This will suffice for our purposes.

*Analysing the data*

Throughout this course, you have investigated and analysed relationships between data, using your graphics calculator. This modern tool was not available to those initially searching for a relationship between the directions of light rays before and after encountering a water surface, so it was several centuries before meaningful results were obtained. With your calculator, you are in a more fortunate position.

*Interpreting the results*

The interpretation needs to explain commonly observed phenomena. For example, when light falls on a flat water surface, such as a pool or lake, a familiar observation is that the light is reflected, at least partially: as a result, you can usually see a reflection of the surroundings in a still pool or lake. You can also often see the bottom of a clear pool (lakes are mostly too deep), so some light must pass through the surface and illuminate the bottom of the pool. Another common phenomenon is that the water in a swimming pool looks shallower than it really is; similarly, a straight stick partially submerged in water appears bent. Hence it seems that light changes direction or is bent as it crosses the boundary between air and water. This behaviour is called *refraction*.

Therefore both the *reflection* and *refraction* of light meeting an air–water boundary need to be investigated. Historically, reflection was investigated much earlier than refraction, and it is the basis of one of the earliest models of the rainbow—that due to Aristotle.

## 2.2 Reflection

When light falls on a flat shiny surface, like a mirror or the surface of a calm lake, it is reflected. There is something very symmetrical about the reflection—it is the same as the object that is being reflected, but back to front, which suggests that the direction of the light rays has changed. In order to express the relationship between the direction of the *incident* (incoming) ray and that of the *reflected* (outgoing) ray, it is necessary to give the angles concerned some names.

Figure 3 shows a common model of reflection. Light rays are represented by straight lines, with arrows indicating the direction of travel of the light; for simplicity, just one ray is shown here.

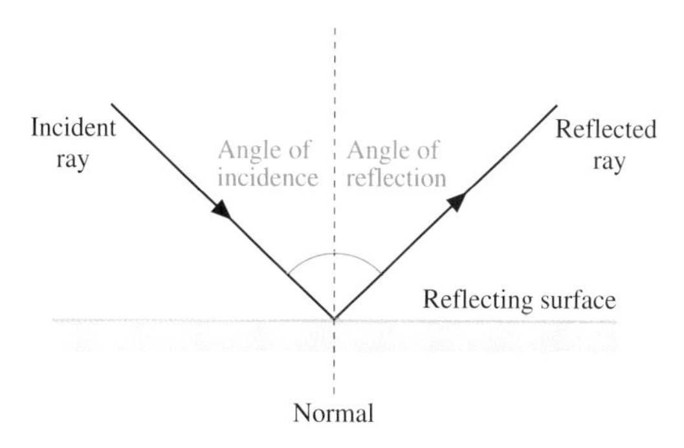


*Figure 3* Representation of the reflection of a ray of light from a flat surface.

The incident ray represents the light striking the reflecting surface, and the reflected ray represents the light reflected from the surface. The *angle of incidence* is the angle between the incident ray and an imaginary line, called the *normal*, drawn at right angles to the reflecting surface at the point where the incident ray strikes the surface. The *angle of reflection* is the angle between that normal and the reflected ray.

For the investigation of reflection, the question posed earlier about the relationship between the directions of light rays before and after encountering a flat air–water boundary can be refined to: What is the relationship between the angle of incidence and the angle of reflection? Data collected on angles of incidence and the corresponding angles of reflection display a very simple relationship: *the angle of incidence is always equal to the angle of reflection*, confirming that reflection has a certain symmetry. This relationship, which was known to the ancient Greeks, is often called the 'law of reflection'.

Note the assumptions of the model illustrated in Figure 3: light is represented by idealized straight lines, and angles are quoted with respect to an imaginary line (the normal) at right angles to the surface. Interpretation of this model should explain, among other things, the symmetry between an object and its reflected image. As Figure 4 shows, when a series of parallel rays of light strike a reflecting surface, the order in which the reflected rays appear is the reverse of the order of the incident rays (the ray represented by a dashed line in Figure 4 is uppermost before reflection, but lowermost after reflection); consequently, there is a reversed symmetry between the object and its image.

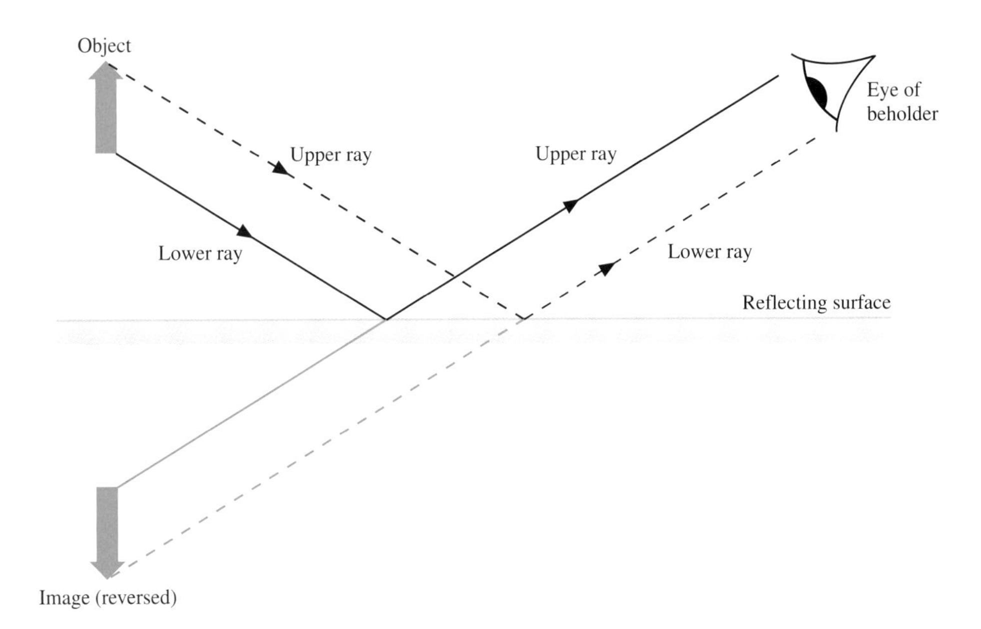


*Figure 4* Reversal of the order of reflected rays of light.

Figures 3 and 4 illustrated reflection from a flat surface, but how does the law of reflection apply to a curved surface, like that of a raindrop? A simple model of a raindrop in two dimensions is a circle (Figure 5). A normal has to be drawn to the surface at the point of incidence, and for a circle, the normal is taken to be an extended radius.

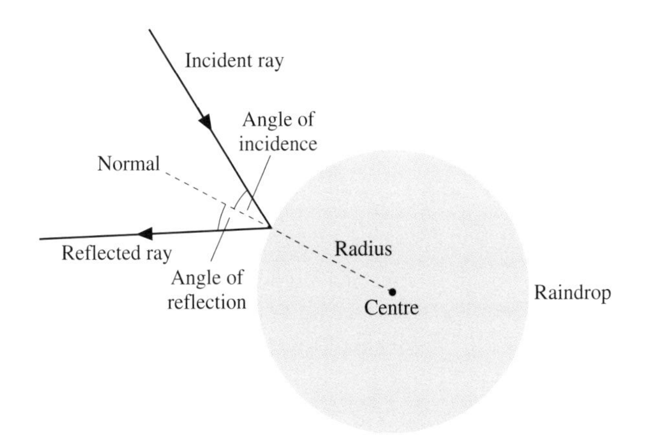


*Figure 5* Reflection of a ray of light from a curved surface, such as that of a raindrop.

As Figure 5 indicates, the law of reflection applies equally to a curved surface: the angle of incidence equals the angle of reflection. However, when parallel rays of light strike a curved surface, the angle of incidence of a particular ray will depend on exactly where that ray is incident on the surface. Therefore reflection from raindrops is not as straightforward as that from a flat surface: the reflected rays seem to be scattered in all directions (see Figure 6).



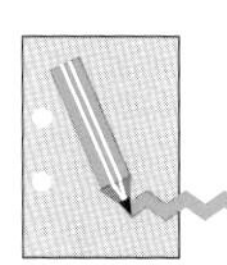

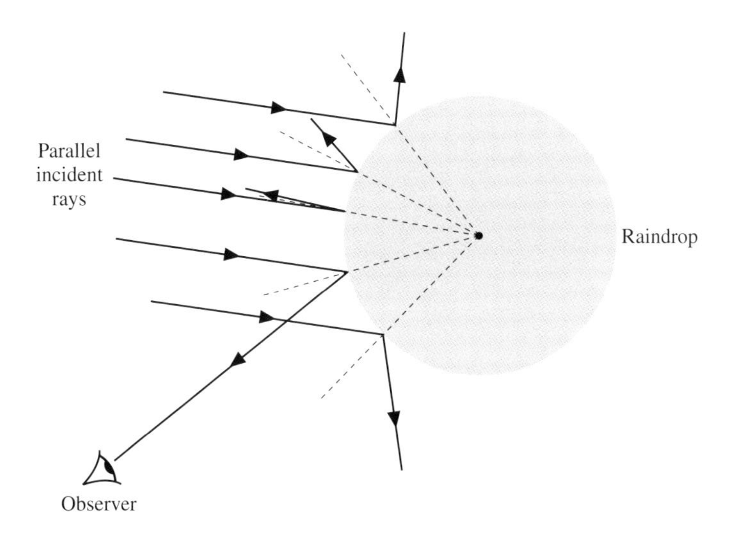


*Figure 6* Reflection of parallel incident rays of light from the surface of a raindrop.

Only some of the scattered rays will reach the eye of an observer. This means that the observer will not see a complete reflected image as with a flat reflecting surface like a lake. Taking into account rays reflected from each of the many raindrops in a cloud adds further complication.

### Activity 6 Handbook activity

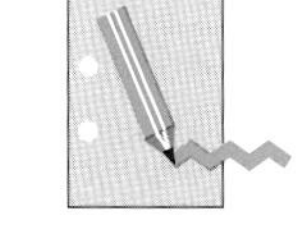

Write some notes about the term 'reflection' on the Handbook activity sheet for *Unit 16*.

## 2.3 An early model of the rainbow

Reflection was the basis of an early mathematical model of the rainbow which is of considerable interest. So it is now time to return to the mathematical modelling cycle.

The Greek philosopher Aristotle (384–322 BC) *created a model* (the second stage of the modelling cycle) in which the rainbow was attributed to a special type of reflection of sunlight from rain clouds. By *doing mathematics* to solve the model (the third stage of the modelling cycle), Aristotle came up with some interesting results. The *interpretation of the results* (the fourth stage of the modelling cycle) is explored on the video sequence for this section.

The early stages of the modelling cycle were discussed in relation to the rainbow in Subsection 2.1.

The underlying idea of Aristotle's model was that when rays of sunlight strike points in a rain cloud they are reflected back at a fixed angle with respect to the inclination of the incoming sunlight. The reflected rays appear strongest when viewed at the fixed angle, so certain parts of the rain cloud look especially bright to an observer: one such bright spot is shown in Figure 7. Taken together, these bright spots constitute the rainbow.

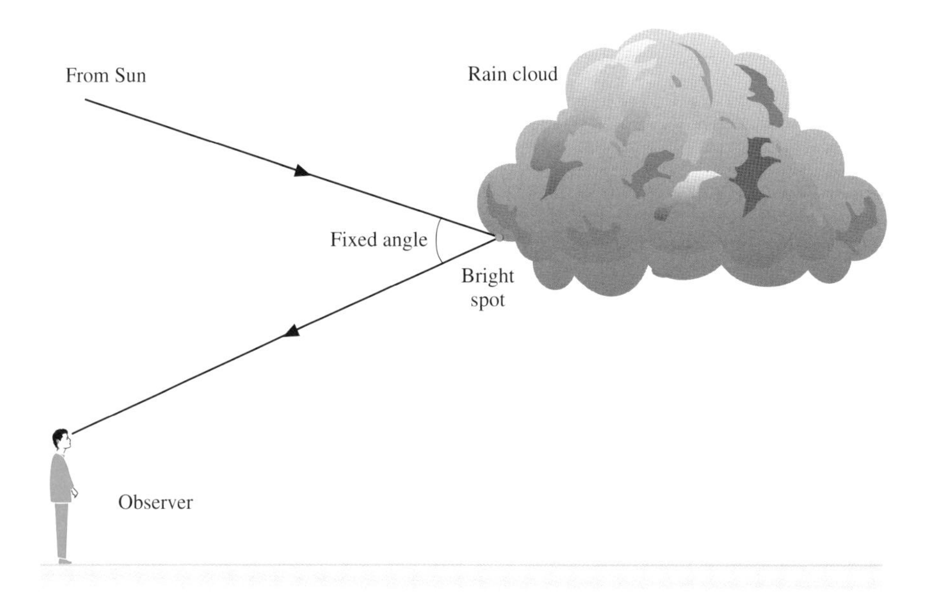


*Figure 7* Aristotle's 'fixed-angle' idea.

The incoming sunlight is incident at different inclinations depending on the time of day. As a result the brightest reflected rays, which would be at a fixed angle with respect to the incident light, would be at different angles *relative to the observer* at different times of day; this explains the varying position of the rainbow in the sky throughout the day (see Figure 8).

So far only two-dimensional representations of the rainbow have been explored. When three dimensions are considered, Aristotle's model also explains the shape of the rainbow (the arc of a circle), as you will see on the video.

*Watch band 13a of DVD00107 now if possible.*

As the video showed, Aristotle's model of the rainbow in three dimensions involves the notion that *every* point in a rain cloud sends back light at a fixed angle *in all directions*—thus each point sends back a cone of light. If you find this difficult to visualize, then you may find it helpful to use an open umbrella as a physical model of what is going on. It is as if light from the Sun comes in along the handle and the shaft of the umbrella, and travels to where the spokes join the shaft. Light then travels out along all the spokes. (This recalls the image at the end of the video section you have just watched.)

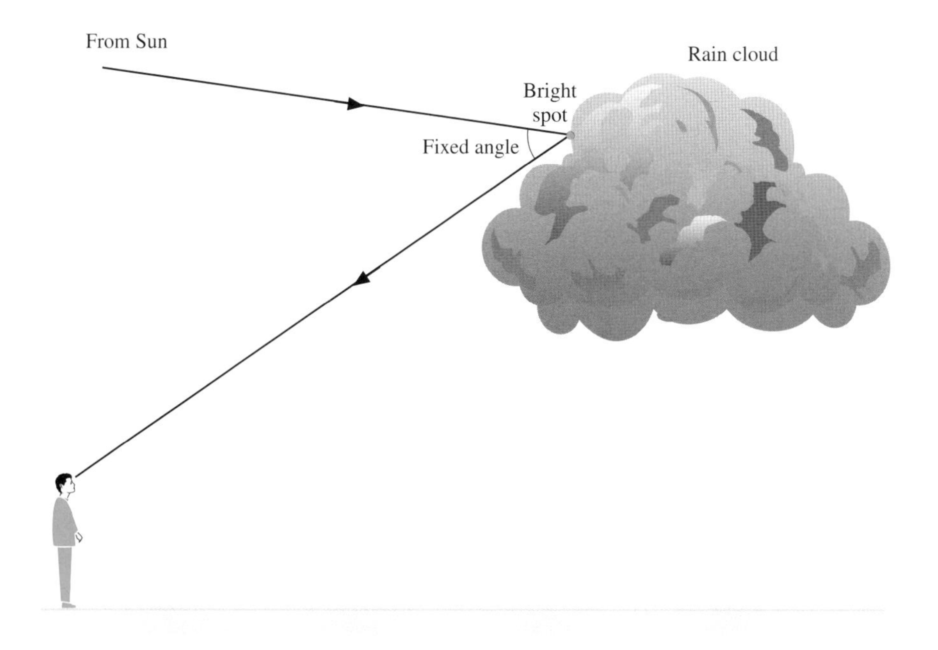


(a) Early morning

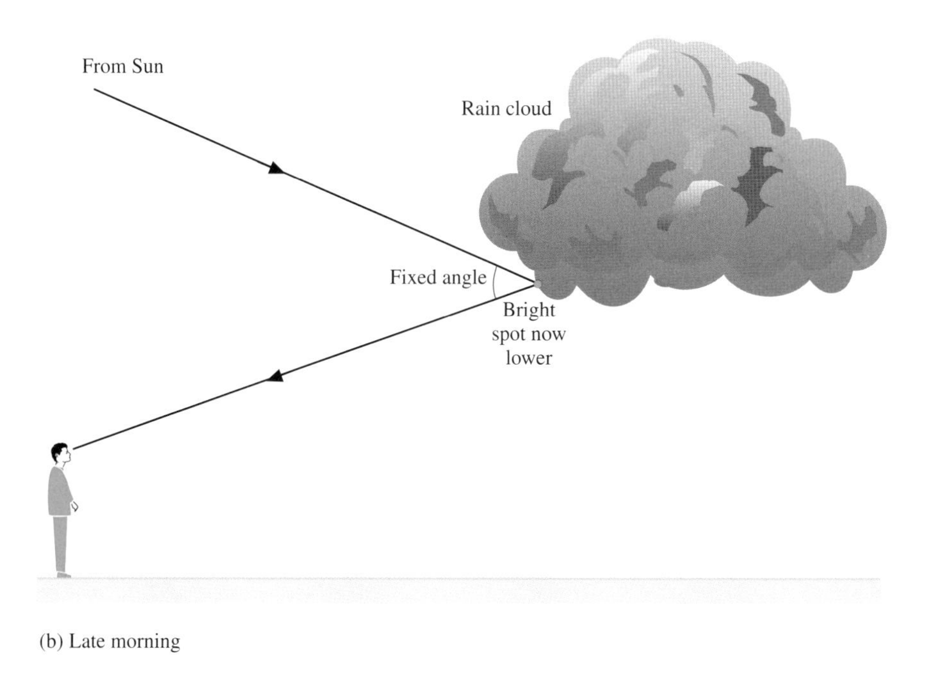


(b) Late morning

*Figure 8* Different times of day—different rainbow positions.

The angle between any spoke and the shaft will represent Aristotle's fixed angle. Of course, an umbrella has only a finite number of spokes, but imagine an umbrella with an infinite number of spokes, with light travelling out along all of them: the light will then form a cone.

The umbrella models an individual point in the cloud, but there are an infinite number of such points, each giving rise to a cone of light. However, only some of the reflected rays of light from these cones are directed to the observer's eye. The geometry is such that the observer sees a circular bow, as in Figure 9.

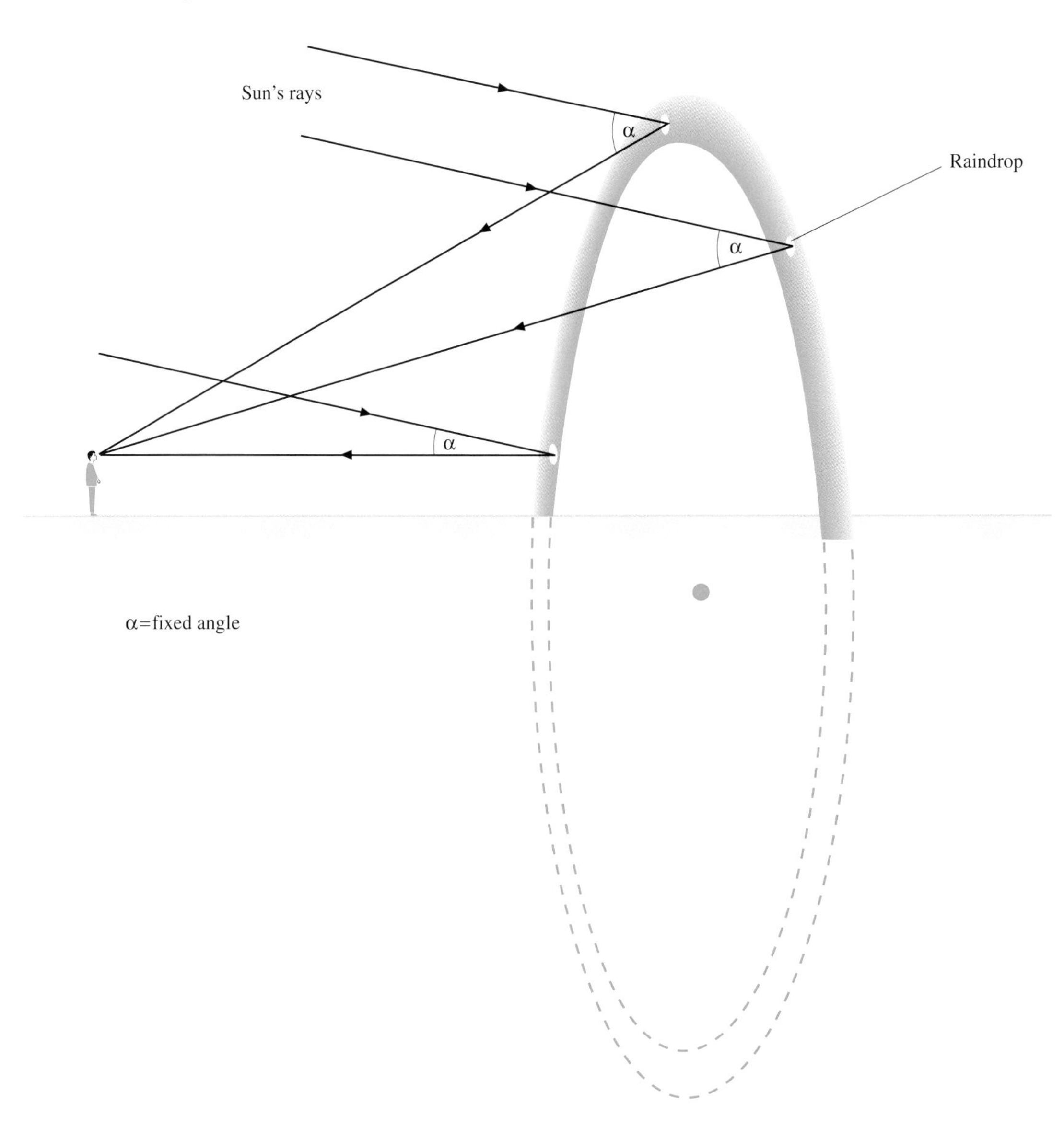


*Figure 9* Aristotle's model in three dimensions: reflections at a fixed angle result in a circular bow.

Thus:

> Aristotle explained correctly the circular shape of the bow and perceived that it is not a material object with a definite location in the sky but rather a set of directions along which light is strongly scattered into the eyes of the observer.
>
> H.Moysé Nussenzveig (1980) 'The theory of the rainbow', in *Light from the Sky*, Freeman, San Francisco, pp.54–65

An interesting feature of Aristotle's model is that it is full of mathematical objects: angles, arcs of circles, cones, lines, a set of directions. These are objects from geometry, a subject with which Aristotle was very familiar.

Now to consider the *evaluation of the model* (the fifth step in the modelling cycle). How good is Aristotle's model at explaining the rainbow phenomenon? It correctly suggests that sunlight has been reflected back to the observer from the rain clouds to give the illusion of an arc of light. Moreover, it accounts for the different positions of the rainbow as the Sun's position changes with the time of day. The model also explains why the end of the rainbow appears to move as you move, just as an object reflected in a mirror appears to move as you move: this means it is impossible to reach the rainbow, just as it is impossible to reach a reflection in a mirror. So, on the basis of the ideas about reflection and the deductive geometric reasoning available to the ancient Greeks, Aristotle's model gives an explanation of the circular shape and the apparent shifting location of the rainbow.

Further supporting evidence for the model came later: measurements made in the thirteenth century showed that the light from a rainbow is concentrated at an angle of about 42° with respect to an imaginary straight line drawn from the Sun (which is behind the observer) to the observer's head (or, more precisely, through the head to the eyes), as shown in Figure 10 (overleaf). Therefore, Aristotle's fixed-angle theory was a good one in some respects.

However, the model had no explanation for the fixed angle itself and why it might be different for different colours, given that the law of reflection is the same for all colours of light. It had a further shortcoming: although the model could account for the fact that the *main* or *primary bow* is at a fixed angle of 42°, it could not explain why there is often another fainter coloured arc, called the *secondary bow*, which appears some 9° higher in the sky, as shown in Figure 10 (overleaf).

Because of these shortcomings, Aristotle's model needed to be modified. Unfortunately, the ancient Greeks did not really understand how light behaves when it passes from air *into* water, as opposed to being reflected from a surface: they did not appreciate the phenomenon of *refraction* which occurs when light goes from one medium to another. It seems that it was not until several centuries after Aristotle that refraction began to be investigated. But the results of these investigations were needed before the rainbow model could be modified successfully by including both refraction and reflection.

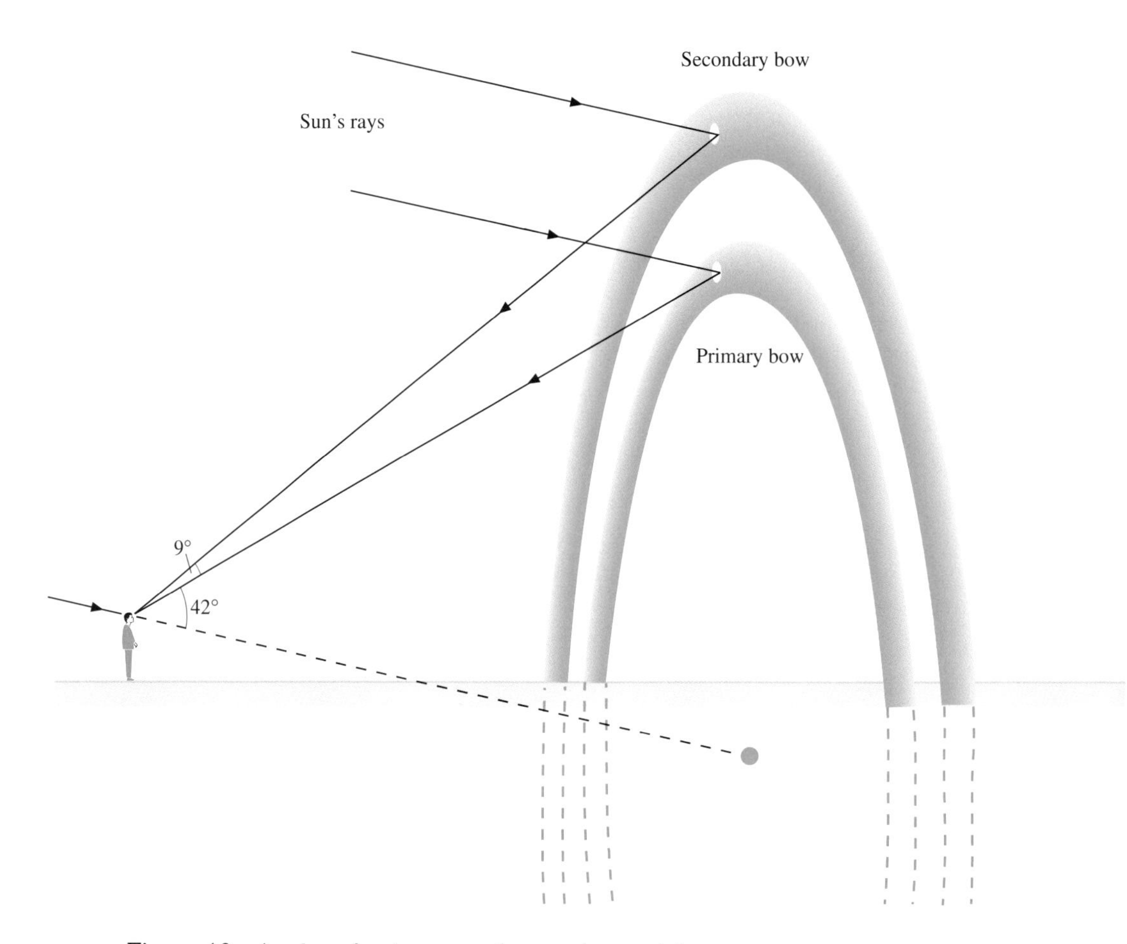


*Figure 10* Angles of primary and secondary rainbows.

## 2.4 Refraction

Figure 11 shows a model of refraction at a flat air–water boundary, like that at the surface of a lake.

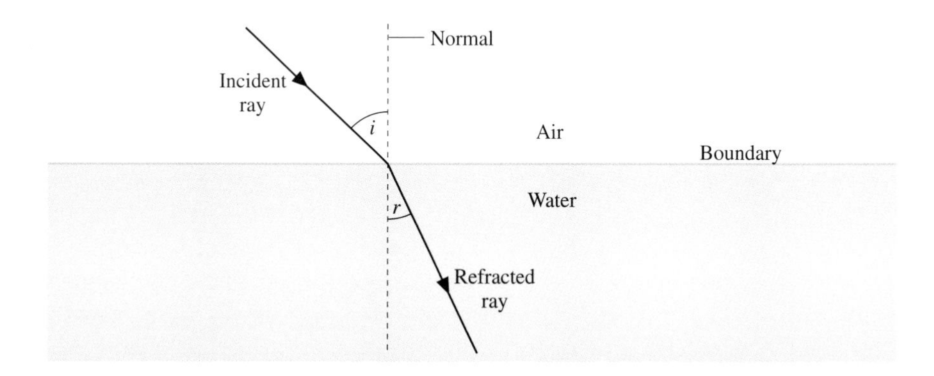


*Figure 11* Representation of the refraction of light at an air–water boundary.

A ray of light meeting the water surface at an *angle of incidence* $i$ (to the normal) is bent (towards the normal) as it enters the water. The *angle of refraction* $r$ is the angle between the normal and the refracted ray in the water. Many generations of scientists have investigated refraction. In terms of the investigation cycle, they posed the question, 'What is the relationship between the angle of incidence and the angle of refraction?' Some of them collected experimental data and tried to analyse the data by fitting different mathematical functions to their findings. The angle of refraction clearly depended upon the angle of incidence: as one increases, so does the other. But it was not a straightforward relationship. Matters were not helped by the fact that the early investigators were attempting the analysis using only pen and paper. You, on the other hand, can use the regression facilities of your calculator to explore the relationship.

## Activity 7 Ptolemy's table

Around AD 100, the mathematician Ptolemy of Alexandria made a number of observations about how the angle of refraction varies with the angle of incidence when a ray of light passes from air into water. His experimental findings are summarized in Table 1.

*Table 1* Ptolemy's values for angles of refraction corresponding to given angles of incidence at an air–water boundary

| Angle of incidence $i$/degrees | Angle of refraction $r$/degrees |
|---|---|
| 0 | 0 |
| 10 | 8 |
| 20 | 15.5 |
| 30 | 22.5 |
| 40 | 29 |
| 50 | 35 |
| 60 | 40.5 |

(a) Enter these data as lists in your calculator. Make a scatterplot of the data on your calculator screen.

What type of function might describe the relationship between the angle of incidence and the angle of refraction?

(b) Use linear regression to predict the angles of refraction that correspond to angles of incidence of 70° and 80°.

(c) Use quadratic regression to predict these two values.

(d) Delete the point $(0, 0)$ from your data lists (to avoid getting an error message) and use power regression to predict the two values.

(e) Discuss which of these models fits the data best, mentioning any other considerations that might be relevant when choosing a regression model. Are there any other regression models for data of this shape?

(f) Later experimental data for $i = 70°$ and $i = 80°$ gave $r = 44.5°$ and $r = 47.5°$, respectively. How well did your models predict these values?

(g) Angles can be measured in radians rather than degrees. How would the regression models have differed if all the angles had been converted into radians?

Ptolemy's experiments were not as accurate as those of later scientists—possibly he saw the quadratic pattern emerging and rounded his data to fit the pattern. Improved experimental data showed that none of the predictions of the models in Activity 7 was very good. So other functions were tried. Data that give a curved plot suggest quadratic or power regression, but a function involving a sine would also be a possibility.

This was the train of thought followed by Johannes Kepler (1571–1630), often considered to be the founder of modern astronomy, in looking for a law of refraction. Although Kepler did not have the benefit of a graphics calculator with regression facilities, he did have better experimental data than Ptolemy.

*Table 2* Kepler's values for angles of refraction corresponding to given angles of incidence at an air–water boundary

| Angle of incidence $i$/degrees | Angle of refraction $r$/degrees |
|---|---|
| 0 | 0 |
| 10 | 7.5 |
| 20 | 15 |
| 30 | 22 |
| 40 | 29 |
| 50 | 35 |
| 60 | 40.5 |
| 70 | 44.5 |
| 80 | 47.5 |

Kepler made many attempts at finding a trigonometric law of refraction. One of his proposed relationships was $r = k \sin i$, where $i$ and $r$ represent the angle of incidence and the angle of refraction, respectively.

## Activity 8 Kepler's proposed relationships

By working through the steps below, check how well relationships of the type $r = k \sin i$ fit the data.

Note that, to be satisfactory, a relationship must work irrespective of whether degrees or radians are used for the experimental data; however, the constants may differ.

(a) Input the data from Table 2 as lists in your calculator. Since, on your calculator, sine regression requires angles to be measured in radians rather than degrees, you should perform list arithmetic to transform the data in Table 2 from degrees to radians (by multiplying by $\pi/180$). Set your calculator to radian mode.

(b) Make a scatterplot of $r$ against $i$ on your calculator.

(c) Use list arithmetic to store $\sin i$ in its own list, and then perform linear regression on the data for $r$ against $\sin i$. Does this support a relationship of the form $r = k \sin i$? What value does linear regression provide for $k$?

(d) You have the advantage of being able to carry out sine regression on your calculator, so you can easily try other functions involving sine, which were too complicated for Kepler to do by hand. Find the best-fit function of the form $r = a\sin(bi + c) + d$ by performing sine regression on the lists for $i$ and $r$ (in radians). Plot this function and comment on how well it fits the data.

---

Kepler correctly thought that refraction should obey a trigonometric law. He suggested a number of possibilities but none was wholly satisfactory, particularly for large angles of incidence. A much better fit to the data was discovered by Willebrord van Roijen Snell of the University of Leiden, Holland, in 1621. The relationship between the angles $i$ and $r$ was, as Kepler had suspected, a trigonometric one. It took the form

$$\sin r = k\sin i,$$

where $k$ is a constant. It is known as Snell's law and is widely applicable.

The same law of refraction was also derived independently by René Descartes (1596–1650), the French mathematician and philosopher, and so in France it is known as Descartes' law.

## Activity 9 Checking Snell's law

Use list arithmetic to calculate and then store $\sin r$ and then use linear regression to check that Snell's law is a good fit to the data in Table 2. What is the value of $k$ that gives the best fit to the data? Does the value of $k$ depend upon whether the data are in degrees or radians?

You already have $\sin i$ from Activity 8.

## Activity 10 Rearranging Snell's law

Snell's law is often expressed in a slightly different form:

$$\frac{\sin i}{\sin r} = n, \qquad (1)$$

where $n$ is the *refractive index* of the media between which the light travels (for example, air–water). Use the algebra you have learned from *Unit 8* onwards to rearrange $\sin r = k\sin i$ into the same form as equation (1) above. Then find the relationship between $k$ and $n$.

Recall that $\sin r$ is a single entity and cannot be split up into sin and $r$ separately (likewise $\sin i$).

---

Experimental data for the angles of incidence and the corresponding angles of refraction can be obtained to a much greater degree of accuracy than given in Table 2. It is also possible to show experimentally that light of different colours is refracted slightly differently and that the values of the refractive index $n$ are slightly different for different colours. Isaac Newton (1642–1727) made several major contributions to optics and, especially, to the theory of colours. It was Newton who discovered that light of different colours was refracted differently. He published his findings in his *Treatise on Optics*; among other things, he stated that there were seven different colours in the rainbow, and he calculated the values of the refractive index for the different colours.

So, to recap, Snell's law is a directly proportional relationship between the sines of the angles of refraction and incidence. The constant of proportionality depends upon the media on either side of the boundary across which the light passes and also upon the colour of the light.

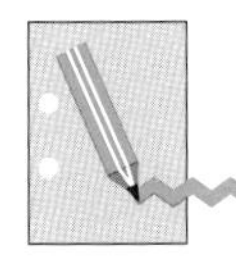

## Activity 11 Revising index

You have met the concept of an index before in *Units 2* and *3*. How does the concept of a refractive index tie in with the idea of an index (for instance, a prices index or an earnings index) as used in these earlier units? What are the similarities and the differences in the usages? How do these meanings compare with other meanings of the word 'index'?

Add some notes on refraction and index to your Handbook activity sheet.

---

The interpretation of Snell's law of refraction explains many everyday phenomena, such as a pool appearing less deep when full of water (see Figure 12(a)). It also predicts that refraction at a water–air boundary is the reverse of that at an air–water boundary. Hence light passing through the parallel boundaries of a rectangular fish tank emerges at the same angle as it entered, although slightly displaced. The same phenomenon occurs when light passes through a block or sheet of glass (see Figure 12(b)); the displacement depends upon the thickness of the glass—the thicker the glass, the more the displacement. Check out this prediction, if you can, by viewing an object (with one eye) through a block or sheet of glass and see if there is an apparent discontinuity in the object at the edge of the glass.

A wave (rather than a straight line) model for light is better for the corners.

Note that the model implied by Snell's law does not explain quite everything that is observed when light encounters a block or sheet of glass, and in particular what happens at the corners. However, since raindrops do not have corners, the model will be very useful for modifying Aristotle's rainbow model.

This modification can now take into account:

◇ reflection (angle of incidence = angle of reflection);

◇ refraction ($\sin r = k \sin i$, with $r$ being the angle of refraction, $i$ the angle of incidence and $k$ a constant for each colour).

Bear in mind that an improved model of the rainbow needs to explain:

◇ the reason for the fixed angle of 42° for the primary bow;

◇ the fact that the fixed angle differs slightly for different colours;

◇ the existence of the secondary bow.

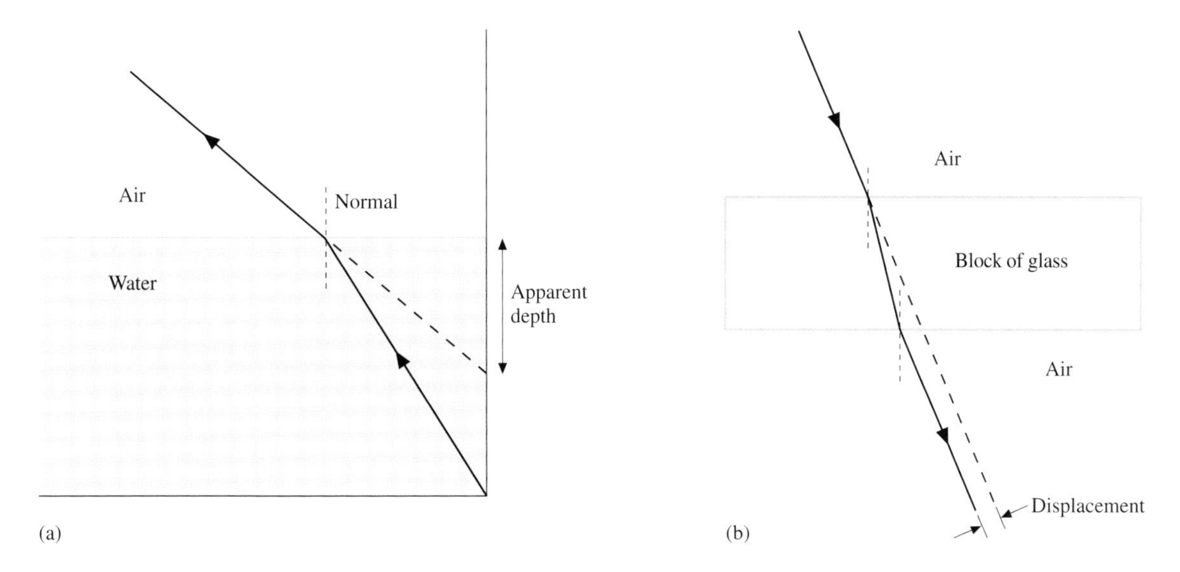


*Figure 12* (a) A pool appears less deep. (b) Light is displaced when passing through a block of glass.

## 2.5 Improving the rainbow model

In a treatise published in 1637, René Descartes offered an improved model of the rainbow. He considered in more detail the role of the individual points in a rain cloud, focusing on the raindrops themselves. His model was based on the idea that each raindrop not only reflects, but also refracts sunlight. Because the refractive index differs slightly for different colours, he thought that refraction might explain why the different colours appear in slightly different positions in the rainbow. The model assumed that when light meets the surface of a raindrop, some of the light is reflected from the surface and some is refracted as it passes through the surface into the water. Similarly, when light meets the water–air boundary at the other side of the raindrop, some light is reflected back into the water, while some is refracted as it emerges into the air.

The rest of this section looks at the development of Descartes' model, using the mathematics that you have met in the course. Fortunately, you will be able to use a graphics calculator, which was not available to Descartes. As you read about the model, try to identify the stages in the mathematical modelling cycle.

An important modelling principle is to start with a model that is as simple as possible and then improve upon it, repeating the modelling cycle with each improvement. This principle applies here. Descartes' model contains the assumption that raindrops are spheres of water; however, first consider a simpler two-dimensional model, and later extend it to three-dimensions (as was done in the case of Aristotle's model). As a further simplification, initially consider light of only one colour, but remember that the value of $k$ in Snell's law will be slightly different for other colours.

So, to begin with, assume that all raindrops can be modelled as identical circles. For simplicity, take the radius of this standard circular raindrop to be one unit (one raindrop radius). Assume that the sunlight falling on the raindrop consists of equally spaced parallel rays of equal intensity. Each individual ray can be specified by its position $X$ relative to a parallel central ray, which passes through the centre of the raindrop (see Figure 13). Because $X$ describes the position at which a light ray *impacts* on the raindrop, it is known as the *impact parameter*.

As Figure 13 shows, the ray of light that impacts on the raindrop, just touching the upper edge of the drop, will be at $X = +1$, and the ray that just touches the lower edge will be at $X = {}^{-}1$. This means that equally spaced parallel rays impacting on the surface of the raindrop will have equally spaced values of $X$ between $+1$ and ${}^{-}1$.

In order to predict what will happen to a light ray when it meets the surface of a raindrop, the angle of incidence is needed. The size of this angle depends on the impact parameter because the raindrop is curved. So it is important to find the relationship between the angle of incidence and the impact parameter. To do this, the trigonometry that you met in *Units 9, 14* and *15* is needed.

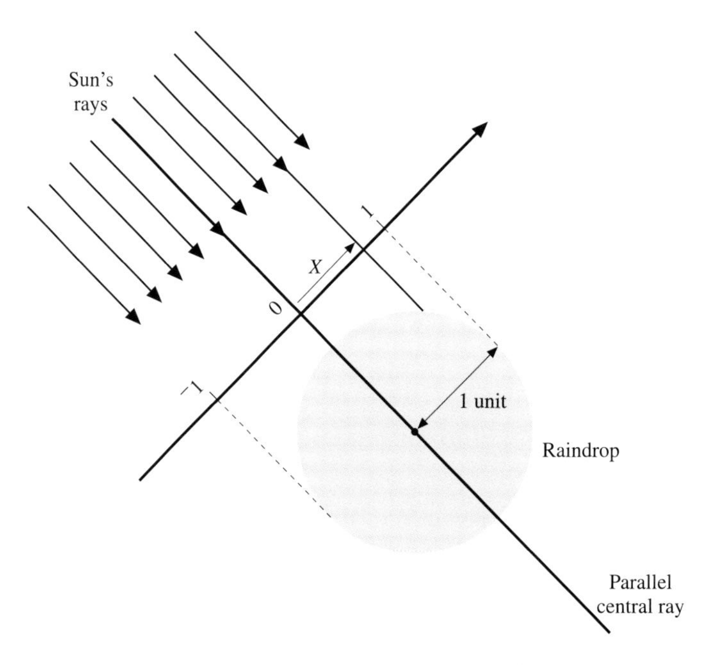


*Figure 13* Model of equally spaced light rays from the Sun falling on a circular raindrop, showing the impact parameter $X$.

## Activity 12 Relating X and i

Use Figure 14 (in particular, triangle $OPQ$) to find the relationship between the impact parameter $X(= PQ)$ and the angle of incidence $i$. Write the relationship in the form $X =$ (a function of $i$). Then rearrange this expression to obtain a formula for $i$ as a function of $X$.

Hint: recall the trigonometric functions and their inverses.

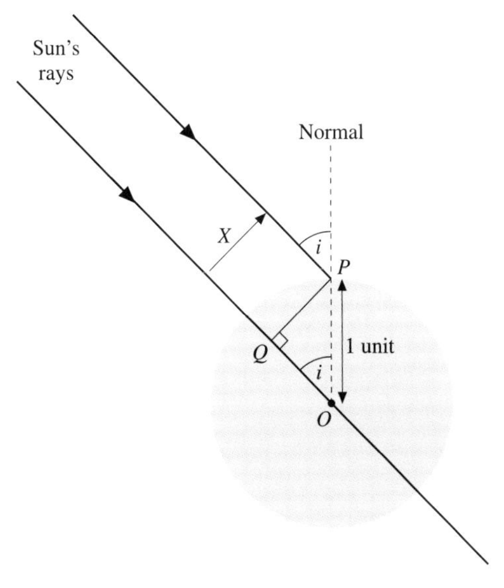


*Figure 14* Relationship between the angle of incidence $i$ and $X$.

Now consider what happens after a ray of light meets a circular raindrop, as in Figure 15. It is reasonable to assume that the light that is *reflected* at the surface of the raindrop is not involved in the formation of a rainbow, as refraction is necessary to produce the different colours.

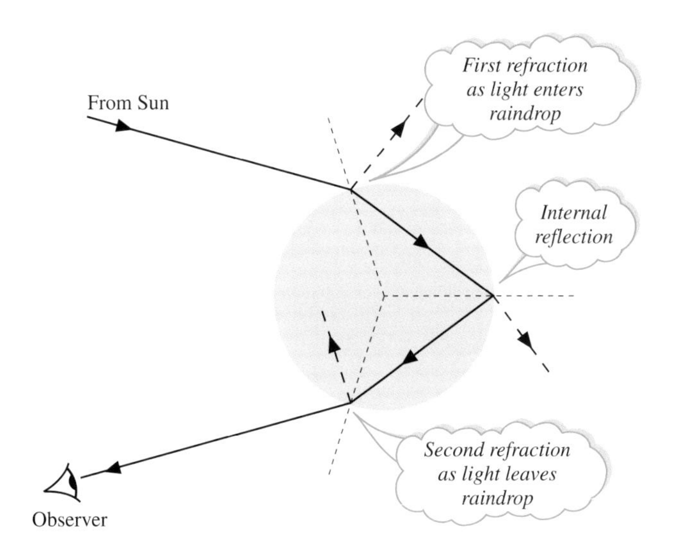


*Figure 15* Path of a ray of light from the Sun through a raindrop to an observer.

Because of this, concentrate on the light that is *refracted* as it enters the raindrop through the air–water boundary. A light ray that is refracted at this boundary will be travelling away from the observer and so will not contribute directly to the rainbow at this stage. But, as Figures 15 and 16 show, the refracted ray next encounters the opposite boundary of the raindrop, and some light is reflected internally from the back of the raindrop. This light then meets the raindrop surface again, where some light will be reflected and some refracted. This refracted light passes out of the raindrop, from the water into air and, if it is travelling in the direction of the observer, it will contribute to the appearance of a rainbow.

The light that is internally reflected again can contribute to a secondary rainbow, as you will see later.

All the *incident* light rays from the Sun that fall on the raindrop are assumed to be parallel and equally spaced. By tracing out the paths of such rays within a circular drop, Descartes showed that the *emerging* rays were neither parallel nor equally spaced, but that they emerged at a range of angles. This tracing of the rays requires a great deal of skill in geometric drawing and is very time-consuming: the angle of incidence has to be determined for each ray and the angle of refraction calculated from Snell's law in order to find the path of the particular ray through the raindrop. This analysis can be done much more rapidly now by using algebra and the graphing facilities of your calculator, but first some geometry is required, using ideas from *Unit 14*.

## Activity 13 Bending sunbeams

Figure 16 represents the path of a ray of light from the Sun as it passes through a circular raindrop.

(a) Consider the light ray that impacts on the raindrop at point $A$. Mark on the figure the angle of incidence $i$ and the angle of refraction $r$.

(b) The direction of the ray changes as it enters the raindrop at $A$—it deviates from its original direction. Mark the angle of deviation on the figure. Write down an expression for the angle of deviation in terms of $i$ and $r$.

Hint: extend the original path of the ray of light, and find the angle between this and the refracted ray's path, in terms of $i$ and $r$.

(c) What happens at point $B$? Look at the triangle $ABO$. What sort of triangle is it? Why is the angle $b$ equal to $r$?

(d) Mark on the figure the angle of deviation of the light ray from its path at $B$. Write down an expression for this angle.

(e) What happens at point $C$? What is the angle $O\hat{C}B$? (Notice the symmetry between point $C$ and point $A$.) Write down an expression in terms of $i$ and $r$ for the angle of deviation of the light ray from its path at $C$.

(f) When the above results (from parts (b), (d) and (e)) are put together, what is the total angle of deviation of the light ray from its original path, as a result of passing through the raindrop?

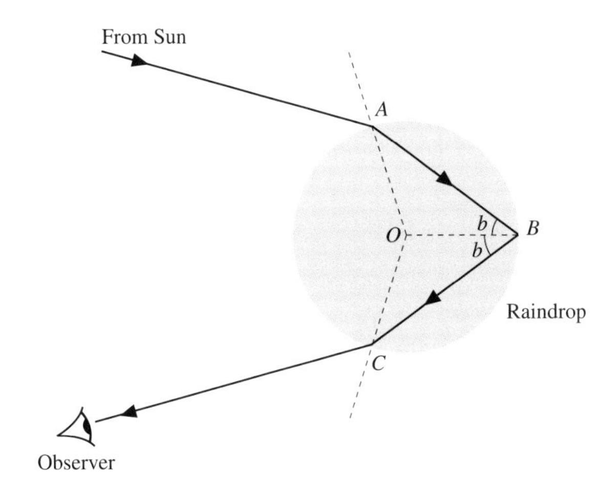


*Figure 16* Path of a ray of light through a circular raindrop.

---

Rather than the angle of deviation, Descartes was interested in calculating the angle *between* the original path of a ray of sunlight on encountering a raindrop *and* the path of the ray on emerging from the raindrop; that is, angle $Y$ in Figure 17 (overleaf). This angle is central to the subsequent discussion of the rainbow model. For convenience, it is sometimes referred to as the *exit angle* or simply as $Y$.

▶ How is the angle $Y$ related to the angle of deviation?

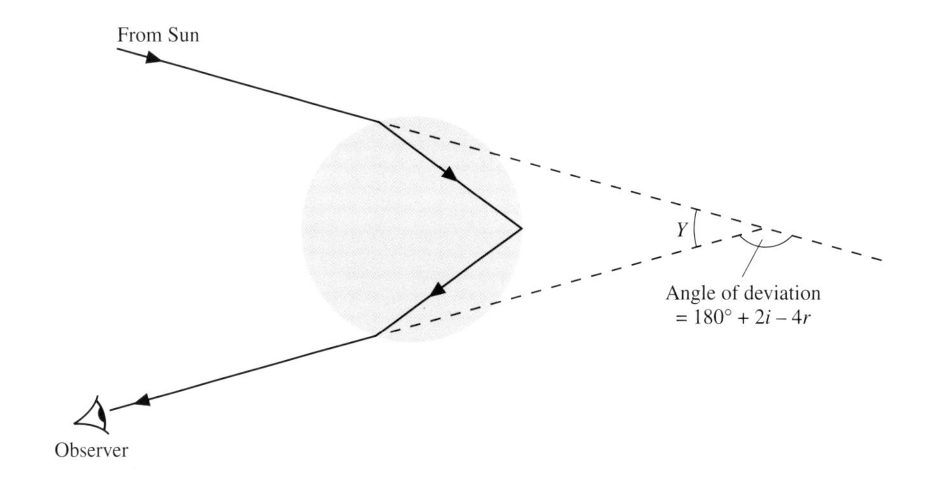


*Figure 17* The deviation of a ray of light on passing through a raindrop.

From Figure 17, you can see that $Y = 180° -$ angle of deviation, so, as the total deviation of a light ray on passing through a raindrop is $180° + 2i - 4r$ (as found in Activity 13), the angle between the ray entering and leaving the raindrop is

$$\begin{aligned} Y &= 180° - (180° + 2i - 4r) \\ &= {}^{-}2i + 4r \\ &= 4r - 2i. \end{aligned} \qquad (2)$$

As you saw in Subsection 2.4, there is a relationship between $i$ and $r$: Snell's law. So, the exit angle $Y$ could be expressed in terms of $i$ alone, but the algebra is easier if the impact parameter $X$ is used rather than the angle of incidence $i$. It turns out that expressing $Y$ in terms of $X$ is more convenient not only for the mathematics but also for the interpretation.

To obtain such an expression for $Y$ you will need to use your algebraic skills, which you have been developing since *Unit 8*.

## Activity 14 Building the model

Recall from Activity 12 that the relationship between the angle of incidence $i$ and the impact parameter $X$ is $X = \sin i$, which can be expressed as $i = \sin^{-1} X$. Also, Snell's law gives $\sin r = k \sin i$.

(a) Use these expressions to write $\sin r$ in terms of $k$ and $X$, and then obtain an expression for $r$ in terms of $k$ and $X$.

(b) Substitute for both $i$ and $r$ in the formula $Y = 4r - 2i$ so as to express $Y$ in terms of $k$ and $X$.

You now have an expression for the exit angle $Y$ (the angle between the path of the incident ray and the path of the ray emerging from the raindrop) in terms of the impact parameter $X$ (a measure of the position at which the ray of light impacts on the raindrop). This expression,

$$Y = 4\sin^{-1}(kX) - 2\sin^{-1} X, \tag{3}$$

will prove very useful in the development of the rainbow model.

It might be helpful here to think briefly about the physical basis for the dependence of $Y$ on $X$. In effect, $Y$ represents the amount of refraction and internal reflection undergone by a given incident ray as it passes through a raindrop. However, the amount of refraction and reflection will depend on the angle of incidence of that ray. In the case of parallel rays of light falling on a raindrop, the angle of incidence of a given ray is determined by the position at which the ray strikes the raindrop—the position being measured by the impact parameter $X$. Hence, $Y$ depends on $X$.

What does the expression for the exit angle (equation (3)) predict in relation to the rainbow?

## Activity 15 The primary bow

(a) Using Descartes' value for the constant $k$, which is $187/250 = 0.748$, and with your calculator in degree mode, plot the graph of $Y$ against $X$ for equation (3), with values of $Y$ from $^-90°$ to $90°$ and $X$ from $^-1$ to $+1$ (recall from Figure 13 that $X$ can only vary from $^-1$ to $+1$).

(b) What are the maximum and minimum values that the exit angle $Y$ can have? Find these values correct to two decimal places.

(c) In the modelling process, the incident rays of light from the Sun have been assumed to be equally spaced and parallel. Use the table facility to look at the values of $Y$ for equally spaced parallel rays from $X = {^-1}$ to $X = +1$, at intervals of 0.1. Is there any bunching in the values of $Y$? Change the table interval to 0.01 and again check if there is any bunching in the values of $Y$.

The results of this activity confirm Descartes' finding that light rays emerge from raindrops at angles of up to about 42° relative to the incident rays from the Sun, with a concentration of rays around 42°, in accord with Aristotle's fixed-angle theory. No light is seen at angles greater than 42°.

For convenience, the angle for the primary bow is rounded to the nearest degree.

The concentration of the emergent rays near 42° means that there will be more light emerging from each raindrop at this angle relative to the Sun's rays than at any other angle. Figure 18 (overleaf) shows that raindrops viewed at an angle of about 42° with respect to the Sun's rays will appear brighter than other drops. By contrast, rays emerging at more than 42° will not be visible to the observer. Something similar happens at $^-42°$, but unless you are on top of a hill or parachuting or in an aeroplane, you will probably not be viewing raindrops at an angle of $^-42°$ relative to the Sun's rays!

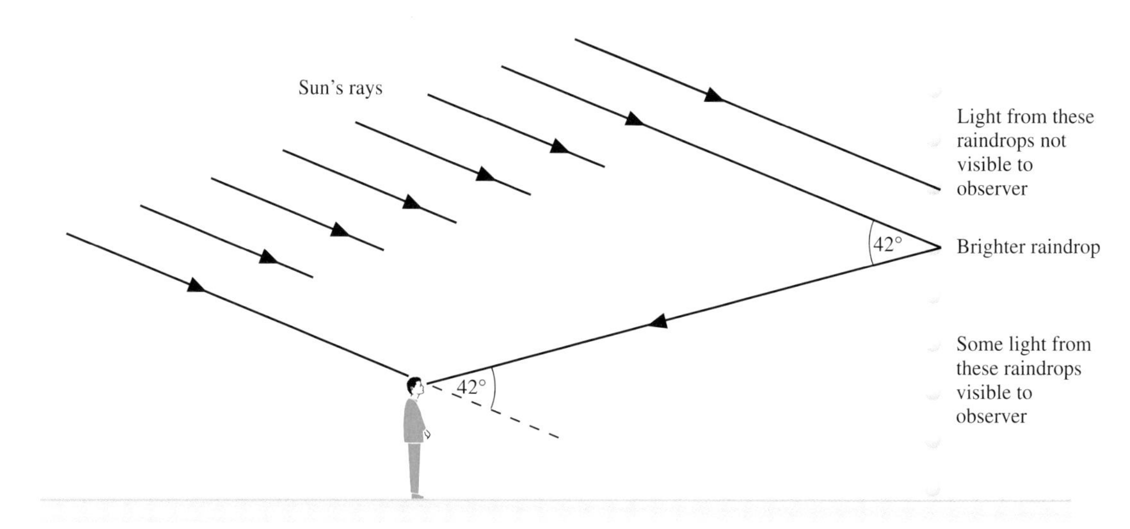


*Figure 18* Brighter raindrops at about 42° relative to the Sun's rays.

So, from a two-dimensional viewpoint, Descartes' model predicts that the rainbow should be concentrated at an angle of about 42° relative to the incoming rays from the Sun. The model is for white light only. However, white light is made up of different colours, and it is split up into these colours by refraction in a raindrop, as shown in Figure 19. The rainbow model can therefore be improved by considering the different coloured components of white light. The value of $k$ is slightly different for each of the colours—Isaac Newton, as a result of experiments, proposed that the value of $k$ for red light passing from air to water was $81/108$, and the value for violet light was $81/109$. Consequently, the brightest light will be at a slightly different angle for each colour.

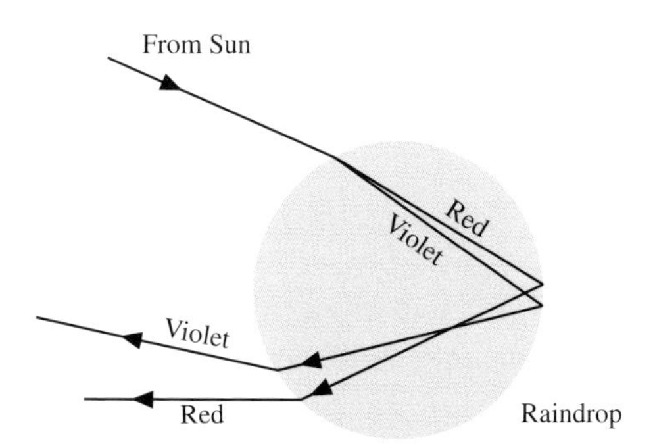


*Figure 19* White light split into colours by refraction in a raindrop.

## *Activity 16* Newton's colours

(a) How do Newton's values of $k$ for red and violet light compare with Descartes' value of $k$ for white light, which is $187/250$?

(b) Using Newton's values of $k$ for red and violet light, in turn, in the function for the exit angle (equation (3)), plot the graphs of the function for both colours. Then find the angles, with respect to the Sun's rays, at which you would expect to see the red band and the violet band in the primary bow.

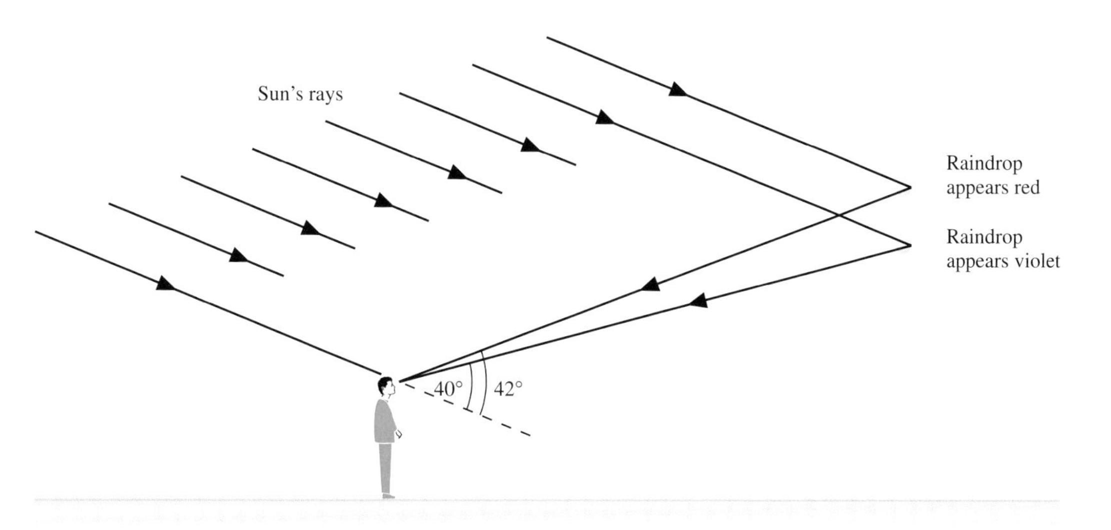


*Figure 20* Different angles for the red and violet bands in the primary bow.

The model has now explained the 'fixed angle' of 42° and why it is slightly different for different colours (see Figure 20). But it has not explained the existence of the secondary bow.

Only light that has been refracted into the raindrop, then internally reflected and refracted out of the raindrop again has been considered so far. Earlier it was mentioned that a second internal reflection was possible, as illustrated in Figure 21 (overleaf). This suggests that the model should be modified to include rays that have undergone a second internal reflection—these might be responsible for the secondary rainbow. Moreover, a second internal reflection would reverse the order of the colours—a feature of the secondary bow. (Remember that reflection reverses the order of the incident rays.)

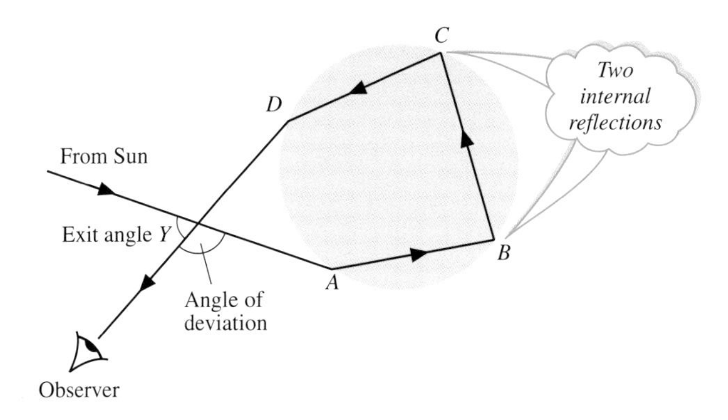


*Figure 21* Path of a ray of light undergoing two internal reflections in a raindrop.

Recall from Activity 13, that one internal reflection (at $B$ in Figure 21) will add $180° - 2r$ to the angle of deviation. So a second internal reflection (at $C$) will add another $180° - 2r$ to the angle of deviation, and correspondingly subtract $180° - 2r$ from the exit angle. Therefore, the formula for the exit angle in the case of two internal reflections is

$$\begin{aligned} Y &= 4r - 2i - (180° - 2r) \\ &= 4r - 2i - 180° + 2r \\ &= {}^{-}180° + 6r - 2i, \end{aligned}$$

but

$$r = \sin^{-1}(kX) \text{ and } i = \sin^{-1} X,$$

so

$$Y = {}^{-}180° + 6\sin^{-1}(kX) - 2\sin^{-1} X.$$

When 360° is added to an angle, it remains the same angle.

It will be convenient to add 360° to the right-hand side of the above equation before plotting the graph, and thereby to express the exit angle for two internal reflections as

$$Y = 180° + 6\sin^{-1}(kX) - 2\sin^{-1} X. \tag{4}$$

## Activity 17 The secondary bow

On your calculator, plot the relationship for the exit angle $Y$ for two internal reflections (equation (4)), with values of $Y$ from 0° to 180° and $X$ from $^{-}1$ to $+1$. Use Descartes' value of $k = 187/250 = 0.748$.

(a) What range of values does $Y$ take when there are two internal reflections?

(b) Find the minimum value that $Y$ takes, correct to two decimal places.

(c) Use the table facility for $X$ between $^{-}1$ and 0 to decide whether there is any bunching in the values of $Y$. If there is, is it as pronounced as that for the primary bow?

(d) What is the angle $Y$ for the secondary bow?

Now consider the interpretation of Descartes' model (in two dimensions). If you combine the graphs from Activities 15 and 17 by plotting the functions for the angles $Y$ for both the primary and secondary bows (equations (3) and (4)) on the same graph, with $X$ between $^{-}1$ and $+1$ and $Y$ between $0°$ and $180°$, then there are a number of interesting features to interpret.

First, notice that the function $Y$ associated with the primary bow is positive only for light rays where $X$ is between 0 and $+1$ (that is, rays incident on what is, effectively, the upper part of the raindrop), whereas the function associated with the secondary bow is positive for $X$ between $^{-}1$ and 0 (that is, rays incident on what is, effectively, the lower part of the raindrop). Second, the function $Y$ for the primary bow takes values up to about 42°, whereas the function for the secondary bow takes values between about 52° and 180°. This leaves a gap.

An interpretation of these findings is illustrated in Figure 22.

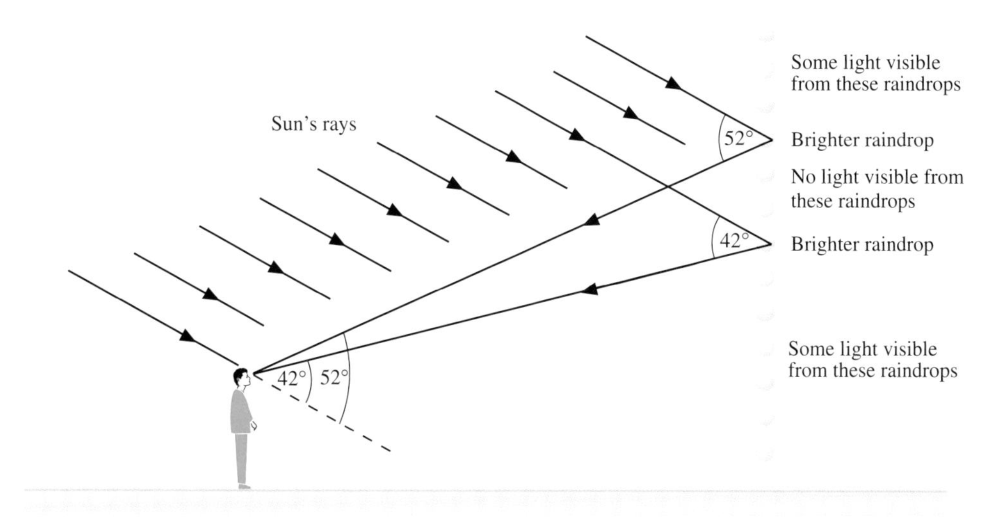


*Figure 22* Interpretation of the two-dimensional version of Descartes' model of the rainbow.

Some light comes from raindrops viewed by an observer at an angle between 0° and about 42° relative to the Sun's rays, with the drops close to 42° appearing brighter. Raindrops viewed at angles between about 42° and 52° will appear dark because neither light from one internal reflection, nor from two internal reflections will emerge between these angles. Raindrops viewed at an angle of 52° or more will be seen to be transmitting some light, with those around 52° appearing a little brighter. Thus, the model predicts a brightness (the primary bow) at about 42° with respect to the Sun's rays, a darkness between about 42° and 52°, and another (less pronounced) brightness around 52° (the secondary bow). The band or darkness between about 42° and 52°, that is, between the primary and secondary bows, is often referred to as *Alexander's dark band* (named after Alexander of Aphrodisias, who discovered the phenomenon around AD 200).

## *Activity 18* Secondary colours

Newton's values of $k$ for red and violet light are, respectively, $81/108$ and $81/109$. As in Activity 16(b) for the *primary* bow, use these values to find the angles, relative to the Sun's rays, at which you would expect to see the red band and the violet band in the *secondary* bow.

A modified model of the rainbow that takes account of Newton's work on refraction can be interpreted as in Figure 23. An observer would see some light coming from raindrops viewed at angles of up to 40° relative to the Sun's rays; between about 40° and 42°, the raindrops would appear to be different colours, with violet at the bottom and red at the top; between about 42° and 51°, the raindrops do not transmit any light to the observer (Alexander's dark band); between about 51° and 54°, the raindrops would appear to be different colours, with red at the bottom and violet at the top; and at angles above 54°, the raindrops would transmit some light to the observer.

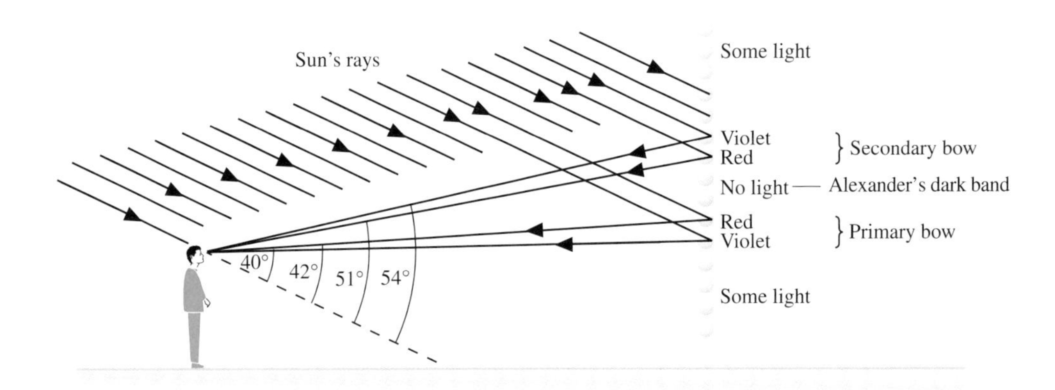


*Figure 23* Interpretation of the modified two-dimensional model of the rainbow.

## 2.6 Modelling the rainbow in three dimensions (video)

The main activity in this subsection is to work through band 13b of DVD00107, which looks at how the mathematical model of the rainbow in three dimensions was modified by the work of Descartes and Newton. The video offers a mathematical account of a number of features of the rainbow—its shape, the order of the colours, and the position of the rainbow in the sky. It also discusses the presence of the secondary bow, and Alexander's dark band between the primary and the secondary bows. It therefore builds upon your work in the preceding subsection.

As you watch the video, bear in mind that the model involves abstract mathematical objects—angles of incidence, of reflection and of refraction, arcs of circles, and cones of light. It uses the language of geometry and trigonometry to describe the behaviour of light. As you follow the model, try to be consciously aware of the representations and abstractions that are being used.

For some people, replacing the rainbow by a collection of mathematical objects may detract from their appreciation of its beauty, but for others the greater understanding brought about by the mathematical model increases appreciation.

> Do not all charms fly
> At the touch of cold philosophy?
> There was an aw[e]ful Rainbow once in heaven:
> We know her woof, her texture; she is given
> In the dull catalogue of common things.
> Philosophy will clip an Angel's wings,
> Conquer all mysteries by rule and line,
> Empty the haunted air and gnomèd mine —
> Unweave a rainbow

(John Keats, *Lamia* II, lines 229–37)

Perhaps the awe of the rainbow will be enhanced for you by the model 'of rule and line'. It is hoped that knowing more about how 'her texture' occurs will not make the 'charms fly' or 'clip an Angel's wings', and that, after studying MU120, you do not feel that mathematics is 'cold philosophy'. Nevertheless, there are undoubtedly people who do agree with Keats' point of view.

*Now watch band 13b of DVD00107.*

The mathematical models of the rainbow developed in this section are useful in understanding most of the phenomena associated with rainbows (see Figures 22, 23 and 24 (overleaf)). But there are a few aspects that they do not explain, such as the supernumerary bands under the rainbow (extra bands of brightness sometimes seen just below the primary bow). To explain these, the model had to be improved further. Instead of light being represented as rays travelling in straight lines, it had to be modelled as a wave. In *Units 9* and *15*, you saw how sound can be modelled as a sine wave and how complex sounds can be thought of as being made up of many different frequency components. In a rainbow, the different colours correspond to light of different frequencies. Thus the rainbow is a manifestation of the frequency spectrum of visible light. In the visible spectrum, red light has the lowest frequency (and the longest wavelength), while violet light has the highest frequency (and the shortest wavelength). If you study optics or the physics of waves, you will learn more about these types of wave.

There are other forms of 'light' with frequencies outside the range that human eyes normally see: for example, infrared light and ultraviolet light.

This section set out not only to 'explain' the rainbow but to raise questions about what modelling a physical phenomenon means. Modelling involves abstraction: seeing raindrops as perfect circles or spheres of water, seeing sunlight as individual rays or as waves, and seeing relationships in terms of geometry and trigonometry.

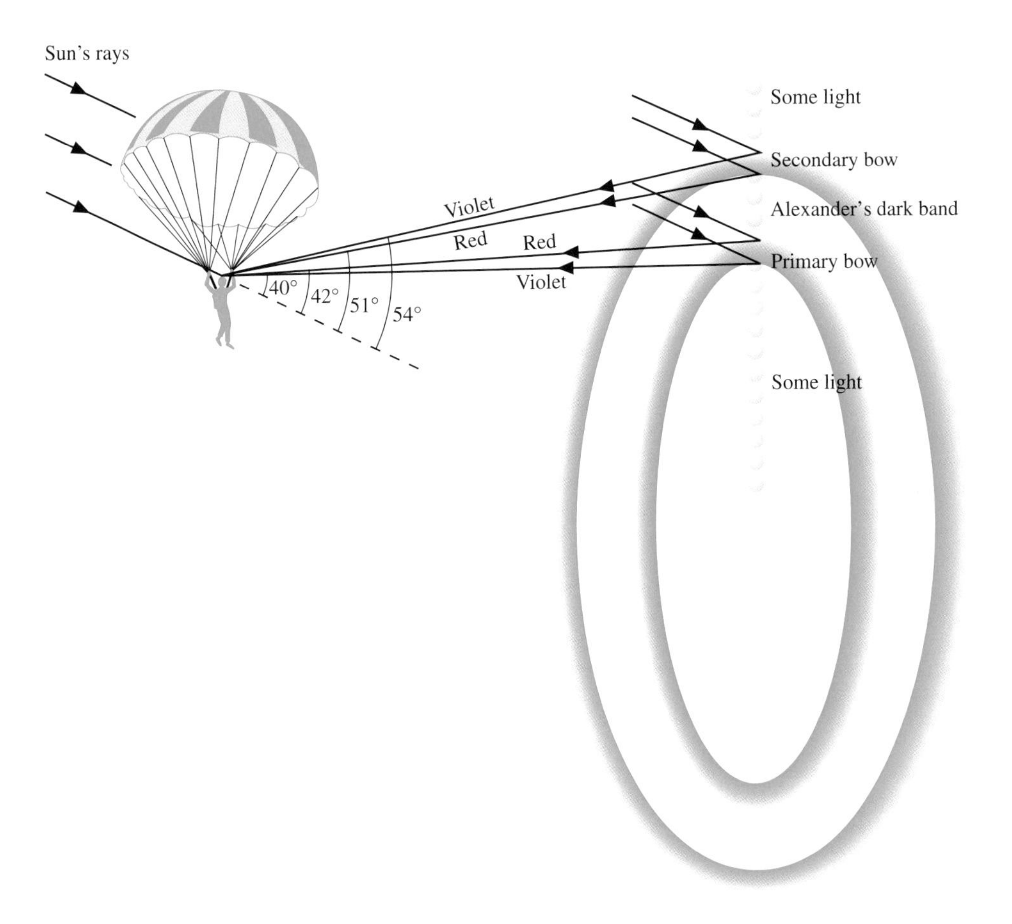


*Figure 24* Interpretation of the three-dimensional 'colour' model, as viewed from an elevated position.

However, within the modelling process, it is *people* who decide which questions are appropriate and which answers are acceptable.

Throughout the course you have been dealing with ways of thinking that are appropriate to a mathematical perspective on the world. The theme of stressing and ignoring weaves in and out of the units. Sometimes it is useful to stress the similarities between the world and its representations; sometimes it is useful to stress the differences. It is this apparent tension which is central to the process of mathematical modelling.

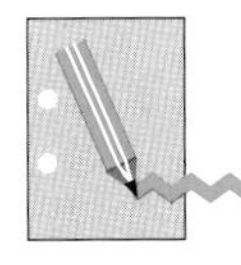

## Activity 19 Rainbow's end?

There is an activity sheet which you may find useful.

(a) Consider the modelling of the rainbow undertaken in this section. Jot down what you now know about mathematical modelling. Compare this with your knowledge at the beginning of the course.

(b) Think about the mathematical techniques you have used while studying the rainbow: data input and regression, geometry, trigonometry, mathematical functions, algebraic manipulation, and graphs.

Write a few notes about your knowledge and skill for each of these techniques: think back to when you started the course and compare what you know now with what you knew in each area at the beginning of the course.

(c) Look at the calculator work that you have done in this section. List briefly the calculator facilities with which you now feel confident. How does this compare with your knowledge and skill at the beginning of the course?

(d) Turn to TMA04 and see if any of the activities in this section are relevant to the questions for Part 1. Would some of your notes for parts (a), (b) and (c) above be of use in answering Part 2? Also glance at the exercises in the Resource Book and see if any of these might be helpful in answering TMA04 or CMA45 questions.

---

You have now revised a number of important mathematical concepts and techniques. The next section concentrates on consolidating your calculator skills, particularly in relation to programming.

## *Outcomes*

After studying this section, you should be able to:

- ◇ explain the terms 'reflection' and 'refraction' as applied to light rays, and calculate angles of deviation and exit angles for light rays that are reflected and refracted (Activities 6, 13);
- ◇ check possible relationships between data by using the regression facilities on your calculator (Activities 7, 8, 9);
- ◇ substitute, rearrange and simplify formulas involving trigometric functions (Activities 10, 12, 14);
- ◇ plot a graph of a mathematical function on your calculator, find a maximum from the graph and/or table, and interpret the results (Activities 15, 16, 17, 18);
- ◇ understand the formation of the primary and secondary bows, calculate the angles between the path of the Sun's rays and the path of the emerging light in the case of both bows (Activities 13, 15, 17), and explain the order of the colours in the bows (Activities 16, 18);
- ◇ relate the mathematical modelling cycle and the statistical investigation cycle to the historical development of models of the rainbow (Activity 19);
- ◇ revise and reflect on your progress in: mathematics; mathematical modelling; and your use of the calculator (Activities 11, 19).

# 3 Creating your own programs

*Aims* The main aim of this section is for you to consolidate your skills in programming the calculator, by studying Chapter 16 of the *Calculator Book*. The section also provides an opportunity for you to review the calculator skills you have acquired during the course. ◇

Previous chapters of the *Calculator Book* have shown you how to create, enter, edit and execute programs and have helped you to build up a small library of programs, stored in the calculator's memory. These programs have involved the use of a broad range of calculator facilities—for example, for dealing with lists or drawing graphs. The programs have been designed to do certain jobs, such as clearing the screen or solving a quadratic equation, but sometimes the task is even more specific and a program has to be written specially.

In the final chapter of the *Calculator Book*, you are encouraged to look back at the programs introduced in the course and take stock of what you have learned about programming. Most of the programming commands are listed and explained, both those you have used before and some new ones, but where necessary, refer to the calculator manual. Think about the various purposes that programs can serve and what makes a 'good calculator program'. The chapter also encourages you to adapt your existing programs and to write others in the future to suit your individual needs. One of these programs might be for TMA04!

You have built up a wide range of calculator skills during the course. Recall the time when you first picked up your calculator and were faced by all those keys, menus and screen displays. You should now know your way round this little machine pretty well, and be aware of when and how it can help you. You have used many of the calculator facilities and you should be confident enough to use the calculator for all sorts of purposes in the future.

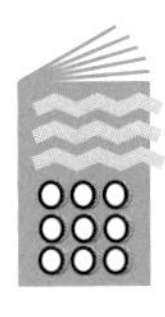

*Now work through Chapter 16 of the Calculator Book.*

Having consolidated your programming skills, and considered the types of program that you might find useful, think about what instructions other people might need if they were to use one of your programs. For instance, some of your programs have been for practising a particular technique, like **POWERS** for multiplying powers; for such programs, you would need to provide instructions on the mathematical techniques involved and on how to use the program. Activities 20 and 21 (overleaf) give you practice in writing such instructions and are very relevant if you are contemplating choosing the calculator question in Part 1 of TMA04. Activity 22 is especially helpful if you choose a calculator-related aspect for Part 2 of

TMA04. If you are not making either of these choices, then you may just like to skim over these activities and read the comments relating to them.

## Activity 20 Program for calculating assessment scores

(a) Think about writing a program to help MU120 students calculate their assessment scores (as in Activity 5). The program will need to ask for the weightings and the assignment scores (you can omit CMA41 as it is zero weighted). Jot down the steps required in the program.

(b) Make a list of the program commands involved and, next to each one, write a comment to indicate its role.

You may want to refer to the calculator manual here.

(c) Write down some brief instructions that will tell someone else how to use the program.

(d) How might the program need to be amended in order to make it suitable for students on other courses with different numbers of assignments and different weightings?

## Activity 21 Programs for other users

(a) Think about writing a program to help another person practise using percentages. The program should generate practice problems of the type 'increase the number ... by ...%'; it should then check the answer and provide an appropriate response. Write down the basic steps that should be included in the program.

(b) What should the user instructions cover?

(c) What additional features might you include in the program to make it easier for others to use?

Look at the calculator question in TMA04 and decide whether you could attempt it. If so, write notes on what will be required for the program and for the user instructions. Check with the comments on Activities 20 and 21 to see if they give you ideas for improvements.

## Activity 22 Calculator progress

(a) Glance back at your answers to Activity 19 (Rainbow's end?) part (c), and your notes on how your calculator skills have developed during the course. Are there other skills you should add to your list from Activity 19(c) in light of your work on this calculator-based section?

You may have used the activity sheet for Activity 19. If so, refer to it now.

(b) Now consider your progress in specific aspects of calculator work, rather than in your calculator skills in general. Focus on one or more of the following calculator skills: handling data; using the statistical facilities; using the graphing facilities; programming the calculator.

If you are choosing a calculator aspect of the course for TMA04, you might like to concentrate on your progress in this aspect. Make notes under headings such as 'My skills now', 'My skills at the beginning of the course', 'My progress during the course', 'Factors that have influenced my progress'.

Look carefully at Part 2 of TMA04 and consider which of your notes are relevant to the given calculator aspect. Remember that you will lose marks for relevancy by including irrelevant material. For example, if the topic is the use of the calculator's graphing facilities, any discussion of your *general* programming skills would be deemed irrelevant; on the other hand, skills concerning the writing of programs that use the graphing facilities could be relevant.

---

## Outcomes

After studying this section, you should have:

◇ reviewed the programs that you have written during the course;

◇ considered how to adapt the programs to suit your own needs;

◇ written a program for another person to use and thought about the user instructions required (Activities 20, 21);

◇ reflected on your progress and achievements in using the calculator (Activity 22).

# 4 A crock of gold

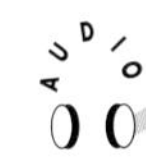

*Aims* The main aims of this section are for you to reflect on what you have learned and how you have learned it during the course, and to consider how the different media have contributed to your learning. ◇

This unit is the end of a journey full of ideas and exploration. Along the way, you have gained knowledge and understanding of mathematics, with your calculator as a useful travelling companion. However, the end of MU120 is unlikely to be the end of your mathematical learning journey, as you are now used to thinking about and 'seeing' mathematics in a wide variety of situations.

The question 'What does it mean to think mathematically?' has run through the course, challenging you to think explicitly about learning and doing mathematics. Many opportunities have been presented to you. You will not have needed to take them all, and not all will have interested you. You will have had your own reasons for studying mathematics and your own expectations about what you should achieve.

Recall the beginning of your journey: in *Unit 1* you listened to the audio band 'Mathematical musings', which encouraged you to begin to look at your surroundings with a mathematical eye. Taking up that theme again—you may be surprised by how much your mathematical perspective has widened. As you listen to the band, note down mathematical ideas from the course that are relevant to each day's musings.

*Now listen to 'Mathematical musings'—band 1 of CDA5508 (Track 1).*

Mathematical musings revisited.

## Activity 23 Back to the beginning

Compare your response to this audio band with that when you listened to 'Mathematical musings' in *Unit 1*. Try to identify what you have learned and how you have learned it.

Think about the mathematical themes that have run through the course—in particular, the ones associated with those you have provisionally chosen for TMA04. Jot down any ideas from the audio band that relate to these themes.

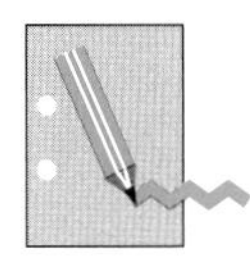

## Activity 24 The medium is the message

There is an activity sheet which you might find useful here.

You have used audio in lots of different learning situations during your study of the course. Write down on the activity sheet the different ways in which audio has contributed to your learning.

Consider the role of video in your learning on the course. Make some brief notes about the ways in which the video in Section 2 of this unit added to your understanding of the rainbow. Are there other ways in which video has played a part in your learning during the course?

Think about other media used in the course: the calculator, the readings, tutorials, and the activities, pictures, diagrams and graphs in the units. How have these contributed to your learning? Are there particular factors that have influenced your progress in certain aspects of your learning?

## Activity 25 Back to the end

Look at Part 2 of TMA04 and at any notes that you made on the different aspects in Activities 2(b), 19, 22 and 24. Are there any additions you want to make to these notes?

At the end of a rainbow, so the saying goes, you should find a crock of gold. The problem is that the end is impossible to find because as you move forward, so does the rainbow. Learning is a bit like that: you never reach the end, no matter how far you go—there is always something new you want to learn. You have probably made progress in many aspects along the way and have found some small crocks of gold. Now, the immediate gold to aim for is completion of the consolidation assignments.

## Outcomes

After studying this section, you should be able to give a reflective account of your progress and achievements on the course. You should also have added to your notes for some parts of TMA04.

# 5 Doing the final TMA

*Aims* The main aim of this section is for you to make a start on TMA04 by identifying and demonstrating what you have learned and how. ◇

In this section, you will work on some TMA04 questions. Have the assignment booklet handy, so you can keep referring to what is required for the assessment. Also have available your past assignments, your learning file and as much of the course material as possible for reference.

Some of the questions in TMA04 require evidence of your work on specific topics from the course. You may already have identified some possible examples to serve as evidence, when you worked through earlier sections of this unit. This section should help you to select the best of these.

If you have chosen the calculator question in Part 1 of TMA04, you should have made a start on it in Section 3. Also if you have enjoyed Units 14 and 15 and Section 2 of this unit, you might have chosen the trigonometry question. However, you will probably need to do at least one consolidation question from Part 1, which requires evidence. There is a reflective question in Part 2 of TMA04 and notes from your work earlier in this unit, especially Section 4, should help you with this.

In order to assist you in starting TMA04, Section 1 suggested some steps to follow. This section looks at these steps in more detail, starting with reviewing mathematical themes.

## 5.1 Reviewing mathematical themes

To help you to focus your review, consider some of the important mathematical themes that have run through the course:

- ◇ numerical relationships,
- ◇ functions and symbols,
- ◇ data and modelling.

These particular themes correspond to some of the questions in TMA04.

You have probably made a preliminary choice of questions. However, it is worth considering each theme in turn before you make your final selection. Think about which topics and which units are relevant to each theme. This review may also help you to identify parts of individual units that might be useful in answering some of the CMA questions. The unit outcomes, the index at the back of each unit and the back of the *Calculator Book* may assist you in identifying relevant work. You may also find the blue Handbook activity sheets a convenient reminder.

Start with the first theme listed: numerical relationships. Initially, consider how this theme cropped up in the first couple of units:

*Unit 1* covered everyday arithmetic, percentage increases and decreases, and powers. These topics also appear in other units later in the course.

*Unit 2* covered the median, mean and weighted mean of a batch of numbers, price ratios, prices index, the relationship between price index and inflation and the purchasing power of the pound. The unit also discussed relative and absolute comparisons between numerical data.

## Activity 26 Numerical relationships

(a) Think back to the other units in Block A. What numerical relationships were covered in these units?

(b) Think about the units in the rest of the course. What other numerical relationships have been covered?

The themes offer a broad view of the course, but to demonstrate your achievements for the purposes of TMA04, you need to focus on specific topics, given in the questions, that relate to the themes. These topics change from year to year. In a previous year, one topic on the numerical relationships theme was ratio. To demonstrate their learning, students were asked to provide an example of their own work involving the concept. Some students chose work relating to price ratios from *Unit 2*. Others chose work on earnings ratios from *Unit 3*; the ratio involved in map scales from *Unit 6*, or in graph scales from *Unit 7*; string-length ratios from *Unit 9*; and trigonometric ratios from *Units 9*, *14* or *15*.

## Activity 27 The numerical relationships topics

Look at Part 1 of TMA04 and the topics in the numerical relationships question. Jot down the different units in which each topic occurs and list the work you have done on the topic in these units.

Table 3 gives topics from a previous year's final TMA for two other themes, together with possible sources of examples on these topics.

*Table 3* Some specific topics within mathematical themes for TMA04 questions

| Theme | Topic | Sources in units |
|---|---|---|
| Functions and symbols | Sine function | *U9* Sine wave representing musical notes<br>*U13* Translation of sine function<br>*U14* Sine as a trigonometric ratio<br>*U15* Phase, amplitude, periods, model of sunrise times, sums of sine functions<br>*U16* Use of sine function in refractive index |
| Data and modelling | Regression models | *U10* Linear regression<br>*U11* Quadratic regression<br>*U12* Exponential regression<br>*U13* Power regression<br>*U15* Sine regression for sunrise times<br>*U16* Investigating refraction data, using regression |

Now that you have a sense of how to review the course material in relation to particular topics which are associated with the themes for the questions, reconsider the questions in TMA04 and finalize your choice.

### *Activity 28* *Other topics in TMA04*

For your chosen questions in Part 1 of TMA04, note the units where there are examples relating to the topics concerned.

The next step is to home in on the specific material that you will use in tackling the topics.

## *5.2 Researching topics*

First, consider what you have learned about a given topic. One way to do this is to identify the relevant learning outcomes from the units. These statements tell you what you should know and what you should be able to do. The section outcomes also give references to related activities. For example, if the topic is *ratio*, you might first look at price ratio in *Unit 2*. Table 4 shows one student's list of learning outcomes relating to price ratios, taken from *Unit 2*. Notice that the student has also identified possible examples, including parts of two assignments.

*Table 4* A student's list of learning outcomes for 'Price ratio' from *Unit 2*

| Learning outcome | Possible examples |
|---|---|
| Convert a percentage price increase into a price ratio or index. | CMA42 Q7–8<br>Section 1 activities |
| Calculate a percentage price increase from a price ratio. | TMA01 Q4 |
| Use price ratios to work out values of a price index. | Section 4 audio |

Note that TMA and CMA questions vary from year to year.

You could go on to do much the same for the other units that include material on ratio. This would give you a good overview of the topic and suggest where to look in more detail later.

Finding the ratio pieces.

## Activity 29 *Identifying learning outcomes*

Identify the relevant learning outcomes for one of the topics in your final choice of questions for Part 1 of TMA04.

You should find that each topic occurs in several units and in various assignment questions. The unit outcomes and the index at the back of each unit and the *Calculator Book* may be helpful here.

You may wish to do the same exercise for the other topics in your chosen questions. At the same time, note down useful references for any questions in CMA45 about which you are unsure.

## 5.3 Selecting evidence of learning

Once you have identified the relevant unit sections for a particular topic, you will need to look more closely at your own work on that topic.

Throughout the course you have demonstrated your progress by answering TMA and CMA questions, completing activities, writing notes, discussing ideas, and so on. Now you need to select evidence of your achievements from the work you have done. This evidence should take the form of written material that you have produced yourself and should demonstrate that it is all your own work. It is important that your evidence shows exactly what is required by the question, and is not just vaguely connected.

Clearly, assignment questions are a major source of evidence. If you use a CMA question as an example, then you will need to include the question and your working for the answer. If necessary, annotate the answer to ensure that it shows your understanding of the topic and was not just copied from the solutions that were sent to you with your results letter. You should also include a copy of your results letter to show that you got the answer right. In the case of TMA questions, your tutor's comments and marks are validation that it is your work. The whole of your TMA answer is unlikely to be relevant. Only select those parts which apply to the topic, and use some method of highlighting or indicating exactly what is applicable. Apart from the assignments and activities, you can draw on work from tutorials or the Resource Books, and/or notes on video and audio materials. Make sure that you include enough of your solution but not too much. Also add an explanatory sentence or two.

Note if you made an error in your TMA answers, you should show corrections that were made in light of your tutor's comments.

The type of evidence that you should choose from your own work on a particular topic will depend upon the specific wording of the TMA04 question. Not only look at parts of the question that ask for examples, but also at parts which ask you to write something about the concept. If you do not have anything to say that will demonstrate your understanding of the topic in relation to your example, look for another.

The work that you present for TMA04 must be focused on the requirements of the questions. For each item you are considering, ask yourself:

◇ Is it relevant to the question?

◇ Is it accurate?

◇ Is it presented so as to show clearly what is relevant to the topic?

◇ Does it demonstrate my mathematical understanding effectively?

Marks will be lost if your work does not meet these criteria, which are the criteria that your tutor will be using. If you think it does not meet some of them, then you may wish to add to it, remove parts of it and/or redo some parts (for instance, if there is an inaccuracy).

If you redo part of a TMA question, include the original as well, so your tutor can verify that it is your work.

## Activity 30 Selecting evidence for TMA04

Find all the examples you can of your own work relating to a topic in one of your chosen questions for Part 1 of TMA04.

(a) Identify examples of your work relevant to the topic.

(b) Note what explanation is required. Then, where relevant, choose the example that is best overall.

(c) Use all the other examples to make notes that might be helpful in the more general explanation required by the question.

## 5.4 Presenting work to show achievements

It is important that your work is presented in such a way that your tutor can see clearly what you understand and can do on each topic. You do not need to rewrite the examples you include but, as mentioned earlier, you may wish to annotate your work in order to highlight particular features or elements. Remember to indicate where each example comes from (for instance, Question 2 of TMA03), as well as which TMA04 question. Give some thought to how you will put the work together, so that your tutor is able to find the way through it easily.

### Activity 31 Organizing your evidence

Gather together all your work for one TMA04 question. Organize and present it clearly to show your understanding of the topics and how the relevant concepts and techniques are used in your selected evidence, as well as in the rest of the course.

If you are doing the calculator programming question, make sure that the program listing and accompanying explanatory comments are easy for somebody else to understand. You may find it helpful to look at the layout and notation in the *Calculator Book*. The instructions for using the program need careful consideration. You might try them out yourself, following each instruction exactly as written, and then perhaps ask somebody else to do the same. In this way you may be able to see the limitations of the program and how it could be improved or adapted. (This might be useful in answering other parts of the question, so do not feel that you have to implement all the possible improvements.) Bear in mind throughout that the program does not need to be complicated—a simple program that does what is required by the question will get full marks, provided all the accompanying documentation satisfies the detailed requirements of the question and is clear and well presented.

## 5.5 Reflecting on progress

As part of TMA04, you need to think back over what you have learned and how you learned it. For example, consider your progress in relation to mathematical modelling. One aspect of Block A involved you using a four-stage cycle for a statistical investigation. In Block B, you were encouraged to think about what was stressed and what was ignored in different representations. In Block C, you met a more detailed modelling cycle and used mathematical functions and regression in modelling. In Block D, you used geometry and trigonometry to develop models. For Part 2 in TMA04, you will need to choose from a number of aspects of your learning and then write a reflective passage on your progress.

## Activity 32 Reflecting on progress

Turn to Part 2 of TMA04 and read over the list of aspects of the course from which you can choose. Identify the ones where you feel you have made progress and, from these, pick out the aspects on which you could write a reasonable amount about your progress (they may not necessarily be the ones in which you feel you have made most progress).

For each possibility, look at relevant activity sheets, notes and assignments from across the course. Then make notes on:

(a) how your approach, skills and/or understanding have changed since the beginning of the course;

(b) your progress in the particular aspect over the whole course, unit by unit or block by block;

(c) the factors that have influenced or changed your approach (for instance, which activities have been most useful in the chosen aspect), along with specific examples of how these factors have affected your progress in the aspect.

Your answers to the above activity should enable you to make your choice from the aspects on offer in Part 2 of TMA04.

Next look more carefully at your notes on the aspect you have chosen and consider what your reflective account might include under the following headings:

- ◇ your skills and knowledge relevant to the aspect at the beginning of the course and now;
- ◇ the way you studied this aspect during the course;
- ◇ the factors that have helped you improve in the chosen aspect, and how they have helped—for example, realizing that you needed more practice;
- ◇ descriptions of the types of activity that you found most useful and why they were useful;
- ◇ activities or components that were *not* useful to you and why.

Although you will not be asked to include examples of your work for Part 2 of TMA04, you should make specific references to appropriate evidence of your achievements in the aspect. Thus, you might refer to skills audits you have completed at various points in the course and quote some relevant comments from them. Feel free to reflect on both good and bad learning experiences.

## 5.6 Completing the final assignments

At this stage you should have some useful notes that can serve as a basis for much of TMA04, and you may have answered a fair proportion of the CMA45 questions too.

In Section 1 you were encouraged to make a plan of your work on this unit and the consolidation assignments. This may be a good time to monitor your progress against that plan.

### Activity 33 What is there still to do?

Look at the assignment booklet containing the final assignments, and also refer back to Activity 4 and your list of things to do. Which things have you completed? Which have you nearly completed? Are there some things that you need to amend (for example, change your choice of questions) or add to your list (items you had previously missed out)? Update the plan and note down what you still need to do to complete the remaining items.

Remember to allow yourself time to check that:

- ◇ the correct number of examples of your work are included—you should not include extra items as they will be considered irrelevant and will lose you marks;
- ◇ all work is clearly headed, and labelled where appropriate;
- ◇ all items are complete and in the correct order.

By now you should have a workable plan for completing TMA04 and CMA45 in plenty of time before the cut-off dates. Your work on this unit and the rest of TMA04 should consolidate your learning and show you just how much you have learned from MU120.

You should find that the skills, ideas and knowledge that you have acquired will be useful in several aspects of your life, and will enhance your understanding of the world.

## Outcomes

After studying this section, you should be able to:

- ◇ carry out a review of your own learning on MU120 by identifying, collecting and selecting evidence of your learning achievements (Activities 26, 29);
- ◇ assemble examples of your work, with relevant explanations, in order to demonstrate your mathematical achievements in specific topics (Activities 27, 28, 29, 30, 31);
- ◇ give a reflective account of your mathematical progress in an aspect of the course, explaining what and how you have learned (Activity 32);
- ◇ plan your work for completing the final assignments (Activity 33).

# Unit summary and outcomes

This unit has focused on consolidating your learning from the course, with the final consolidation assignments in mind.

Section 1 asked you to consider what was involved in tackling the consolidation assignments—both the CMA and the TMA. It encouraged you to think about your strategy for completing the assignments.

Section 2 revised much of the mathematics and calculator work from the course by looking at the mathematical modelling of the rainbow. You saw how an explanation of the appearance and position of a rainbow came from mathematical models developed by a process of abstraction: seeing raindrops as perfect circles or spheres, treating reflection and refraction geometrically, regarding light as straight rays. You saw how the modelling cycle was repeated, improving the model each time. You saw that it is often preferable to start with a simpler model, for instance, a two-dimensional version and then extend it to three-dimensions as shown on the video. You also carried out a statistical investigation of data on refraction and used the regression facilities on your graphics calculator to find a relationship that had taken centuries for eminent scientists without calculators to establish.

Section 3 was based on Chapter 16 of the *Calculator Book*, and reviewed the programming commands and features of your calculator. The chapter included suggestions about how you might organize the library of programs you have entered during the course, and offered pointers about how you might modify and improve them for your own use.

In Section 4 you revisited the audio band on mathematical musings from *Unit 1* and considered how far you have come since you first listened to it. From relationships among numbers in Block A to the rainbow in Block D, you have come a long way.

Lastly, Section 5 concentrated on identifying evidence of your achievements and your progress for use in TMA04. It suggested how you might collect, select and reflect upon examples of your work as evidence of your learning in particular topics.

This unit has encouraged you to take an overview of your progress and achievement over the whole course. Now you should be well prepared to implement the rest of the plan you developed in Activities 3 and 4 in Section 1 for completing CMA45 and TMA04.

Remember, from Section 2, that different people see their own individual rainbows. It is hoped that this course has helped you to see and enjoy your own mathematical rainbow. On your continuing mathematical learning journey, may you find crocks of gold to help you on your way. Good luck!

UNIT 16
UNIT 14
UNIT 15
UNIT 13
UNIT 8
x = ?
UNIT 11
UNIT 9
UNIT 10
PREDICT YOUR COURSE SCORE
UNIT 7
UNIT 5
UNIT 12
UNIT 4
UNIT 2
ENTRANCE
£1
YOU ARE HERE
UNIT 6
UNIT 3
UNIT 1 MATHEMATICS EVERYWHERE

## Outcomes

When you have studied this unit, you should be able to:

- ◇ plan your work for the final assignments;
- ◇ describe the processes of mathematical modelling and statistical investigation;
- ◇ check possible relationships between data by using the calculator's regression facilities;
- ◇ explain the terms 'reflection' and 'refraction' as applied to light rays, and calculate angles of deviation and exit angles for light rays;
- ◇ understand the abstractions involved in Aristotle's and Descartes' models of the rainbow, and also their limitations;
- ◇ explain the formation of the primary and secondary bows of a rainbow, and account for the angles at which they appear relative to the Sun's rays;
- ◇ explain the order of the colours in the primary and secondary bows;
- ◇ review the programs you have written during the course;
- ◇ consider how to adapt the programs to suit your own needs;
- ◇ carry out a review of your learning by identifying, collecting and selecting evidence of your learning achievements;
- ◇ give a reflective account of your mathematical progress in an aspect of the course, explaining what and how you have learned.

# Comments on Activities

## Activities 1–4

There are no comments on these activities, as your CMA strategy, your choice of questions etc. will be personal to you.

## Activity 5

You can check your result on the program on your student home page. Here is an example of one student's scores in a previous year. Note the weighting of the assignments may vary from year to year.

| Assignment | $w$ | $x$ | $wx$ |
|---|---|---|---|
| CMA42 | 7 | 80 | 560 |
| CMA43 | 7 | 50 | 350 |
| CMA44 | 7 | 40 | 280 |
| CMA45 | 14 | 0 | 0 |
| TMA01 | 13 | 80 | 1040 |
| TMA02 | 13 | 60 | 780 |
| TMA03 | 13 | 50 | 650 |
| TMA04 | 26 | 0 | 0 |
| Total | 100 | – | 3660 |

$$\frac{\sum wx}{\sum w} = \frac{3660}{100} = 36.60.$$

So this student's assessment score so far is 37%. This is not far off a pass, even before adding in the scores for CMA45 and TMA04. However note that the student must also achieve the threshold in TMA04.

## Activity 6

You should mention that the angle of reflection equals the angle of incidence, and that the order of light rays is reversed by reflection. Drawing diagrams like Figures 3 and 4 might be useful.

## Activity 7

(a) Since the data pass through the origin, a number of different functions might fit the data quite well. Possibilities are: linear, quadratic, power or, even, a sine function.

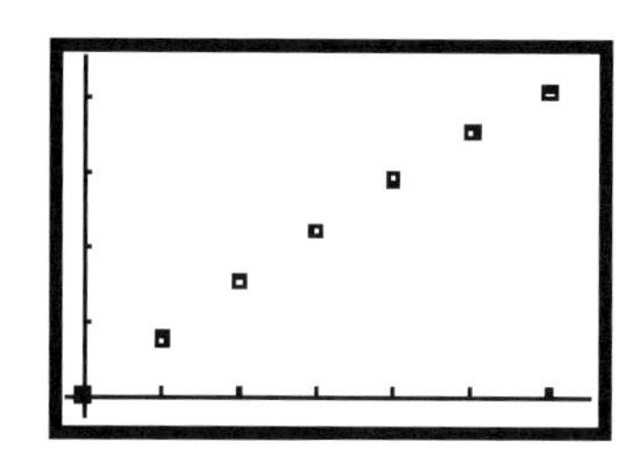

Note that exponential and logarithmic functions do not pass through the origin and so are not likely to be appropriate. The scatterplot suggests a very slight curve, and so as well as linear, quadratic or power regression look possible.

(b) Linear regression gives

$$Y = 0.675X + 1.25,$$

where $X$ is the angle of incidence and $Y$ is the angle of refraction. The correlation coefficient is 0.9979 (to 4 d.p). So this is a good fit to the data.

This model predicts 48.5° and 55.25° for the angles of refraction that correspond to angles of incidence of 70° and 80°.

(c) Quadratic regression gives

$$Y = {}^{-}0.0025X^2 + 0.825X.$$

No correlation coefficient is obtained with quadratic regression, but the value of $R^2$ is 1, indicating a perfect fit. This is confirmed by plotting the regression function on the same graph as the data.

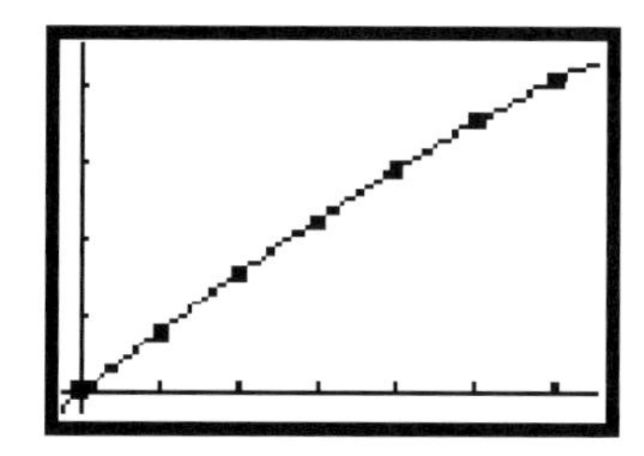

This model predicts 45.5° and 50° for the angles of refraction that correspond to angles of incidence of 70° and 80°.

(d) Power regression (omitting the origin from the data) gives

$$Y = 1.007X^{0.908},$$

with the coefficients given correct to 3 d.p. The correlation coefficient is 0.9996 (to 4 d.p). This model fits the data well, but plotting the function on the same graph as the data suggests that the fit may be less satisfactory at the higher points.

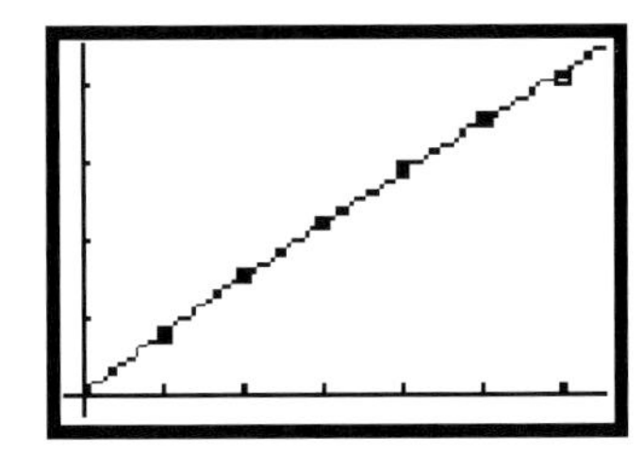

It predicts 47.7° and 53.8° for the angles of refraction that correspond to angles of incidence of 70° and 80°.

(e) The graphs of the different regression functions on the scatterplot show that the quadratic regression function is the best. However, this is a 'sad' parabola and so there will be a maximum, which might not be appropriate for the data. Another possible regression function that should be considered is sine regression.

(f) None of the models did very well.

(g) If the angles had been measured in radians, then the 'shape' of the data would remain the same (everything would be scaled by $\pi/180$). The parameters in the various regression functions would be different, but the goodness-of-fit would be exactly the same.

## Activity 8

(a) Multiply each list by $\pi/180$ to convert from degrees to radians, and then store it as a new list.

(b)

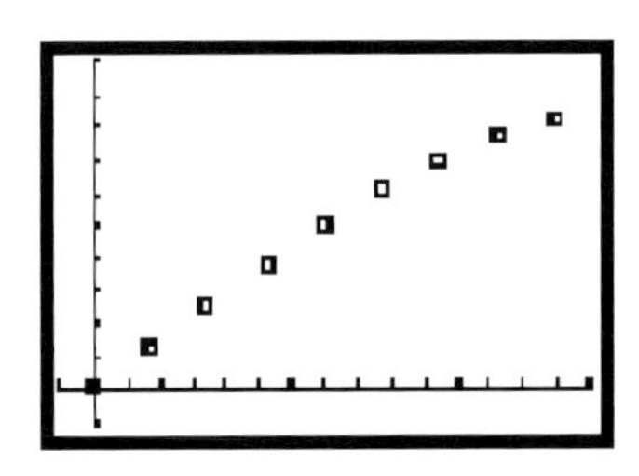

(c) Linear regression on the $r$ and $\sin i$ lists gives

$$r = 0.84 \sin i - 0.018,$$

with the parameters given correct to 2 s.f. The correlation coefficient is 0.9989. So this is a very good fit to the data. However, the $^{-}0.018$ term is a bit unsatisfactory, suggesting that the function does not quite go through the origin, as would be expected from the data.

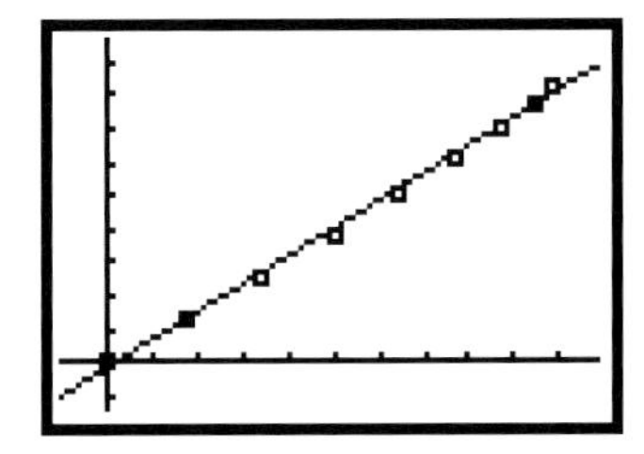

The best value of $k$ would be 0.84.

(d) Sine regression on the $i$ and $r$ lists gives

$$r = 0.707 \sin(1.06i - 0.224) + 0.158,$$

with the parameters given correct to 3 s.f.

This a bit different in form from Kepler's proposed type of relationship. No correlation coefficient is obtained, but the graph shows that the function fits the data very well. However, it does not give good results for negative values of $i$, which should mirror those for the positive values of $i$. So this form of sine model can still be improved upon.

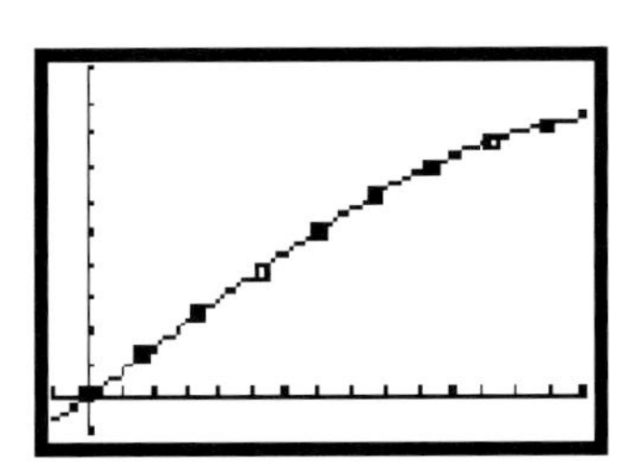

## Activity 9

With the angles in either *degrees* or *radians*, linear regression gives

$$\sin r = 0.747 \sin i + 0.00139.$$

The correlation coefficient is 0.999979, which is the best so far. However, there is still the worry of a small positive constant: +0.00139. But with more accurate data than Kepler's, this constant disappears.

The best value of $k$ is 0.747 (to 3 d.p.).

As linear regression shows, $k$ does not depend upon whether the angles are in degrees or radians, and this is a great improvement upon the other relationships tested.

## Activity 10

From

$$\sin r = k \sin i,$$

you want to obtain an expression for $\sin i / \sin r$. So first divide both sides by $k$ to get

$$\frac{1}{k} \sin r = \sin i,$$

then divide both sides by $\sin r$ to get

$$\frac{1}{k} = \frac{\sin i}{\sin r}.$$

So, in the form of equation (1),

$$\frac{\sin i}{\sin r} = \frac{1}{k}.$$

Hence $n = 1/k$.

## Activity 11

A prices index is based upon a *ratio* of prices (of a basket of goods) at different times. An earnings index is based upon a *ratio* of (average) earnings at different times. The refractive index is based upon the *ratio* of the sines of the angles that indicate the directions of light in different media.

Hence all three types of index are based upon a ratio—that is, a relative comparison, as opposed to an absolute comparison (subtracting the two values). The units used do not affect the value of the index for prices, earnings e.g. pounds or pennies) nor do they for the refractive index (e.g. degrees and radians).

However, the prices and earnings indices are specified with respect to a base date (for example, January 1984) in order to show how the indices change with time. This is not necessary for the refractive index, which does not change with time but varies with the colour of the light and the media through which the light travels.

There are a number of meanings for the word 'index', one being a pointer or indicator. This could be interpreted to cover the above mathematical meanings, as well as the index at the back of this unit and the (index) finger often used to point at it.

An index can also be used to refer to the power to which a number is raised; for example, in $12^4$, the index is 4.

## Activity 12

In triangle $OPQ$ in Figure 14, the angle $P\hat{O}Q$ is equal to $i$. Since the triangle has a right angle at $Q$, trigonometric ratios can be used directly on the triangle. The radius of the circle representing the raindrop is 1, so $OP = 1$ and the side $PQ = X$. These sides are, respectively, the hypotenuse and the opposite side to the angle $i$. Therefore, the sine ratio is the appropriate relationship to use here:

$$\sin i = \frac{X}{1} = X.$$

Hence $X = \sin i$.

In order to find the angle $i$, you need to use the inverse function of sin, that is, $\sin^{-1}$. So

$$i = \sin^{-1} X.$$

## Activity 13

(a) The angles of incidence and refraction are marked on Figure 25.

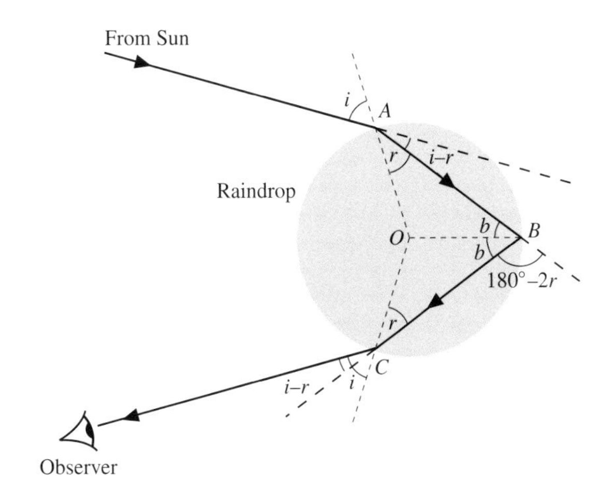


*Figure 25*

(b) At point $A$, the light ray enters the raindrop and is refracted, being bent towards the normal. It deviates from its original path by an angle of $i - r$.

(c) At point $B$, the ray is internally reflected, so the angle of incidence is equal to the angle of reflection.

The triangle $ABO$ is an isosceles triangle, as $AO = BO = 1$ (the radius of the circle representing the raindrop is 1). Hence the angles $b$ and $r$ are equal.

(d) At $B$, the angle of deviation of the light ray from its path is $180° - 2b$, and since $b = r$, this is $180° - 2r$.

(e) At $C$, the ray is refracted away from the normal. As $OBC$ is an isosceles triangle, $O\hat{C}B$ is $r$. Therefore, the angle of deviation at $C$ is $i - r$ (the same as at $A$).

(f) The total angle of deviation of the light ray is the sum of the deviations at $A$, $B$ and $C$ or

$$(i - r) + (180° - 2r) + (i - r) =$$
$$180° + 2i - 4r.$$

## Activity 14

(a) Substituting $\sin i = X$ into $\sin r = k \sin i$ gives $\sin r = kX$. Hence $r = \sin^{-1}(kX)$.

(b) Substituting $i = \sin^{-1} X$ and $r = \sin^{-1}(kX)$ into $Y = 4r - 2i$ gives

$$Y = 4\sin^{-1}(kX) - 2\sin^{-1} X.$$

## Activity 15

(a)

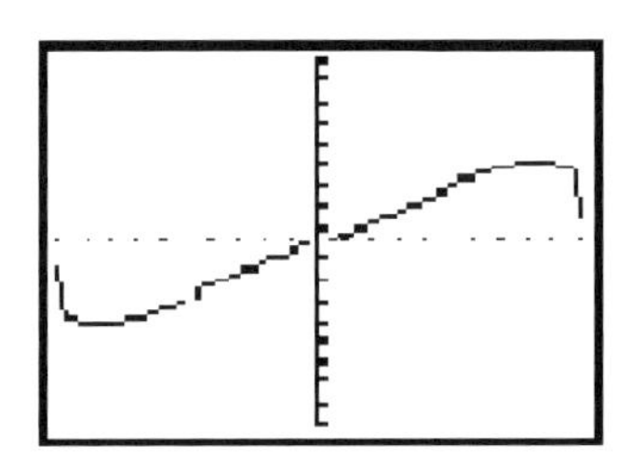

(b) You may have found these values from the graph or table, or by another method.

The maximum value of $Y$ is 41.51° and the minimum value is $^{-}41.51°$. These correspond to values of $X$ of about 0.86 and $^{-}0.86$.

(c) The difference between the values of $Y$ for successive positive values of $X$ (at intervals of 0.1) gets smaller up to $X = 0.9$. It then increases between $X = 0.9$ and $X = 1$. This corresponds to bunching in the values of $Y$ that are just above 41°. A similar effect is seen for negative values of $X$, with bunching in the values of $Y$ that are below $^{-}41°$.

When you look at a table interval of 0.01, the effect is even more pronounced. There is bunching in the $Y$ values around the highest value of $Y$, 41.51°, and the lowest value, $^{-}41.51°$.

## Activity 16

(a) According to Newton, for red light, $k = 81/108 = 0.75$, and for violet light, $k = 81/109 = 0.7431192661$. Descartes' value of $k$ for white light is $187/250 = 0.748$, and so it lies between Newton's values.

(b) The angles at which you would expect to see the red and violet bands in the primary bow are found by looking at the relevant graph or table to obtain the maximum values of the function

$$Y = 4\sin^{-1}(kX) - 2\sin^{-1} X,$$

with the appropriate values of $k$. The angles for the red and violet bands turn out to be about 42.0° and 40.3°, respectively.

## Activity 17

(a)

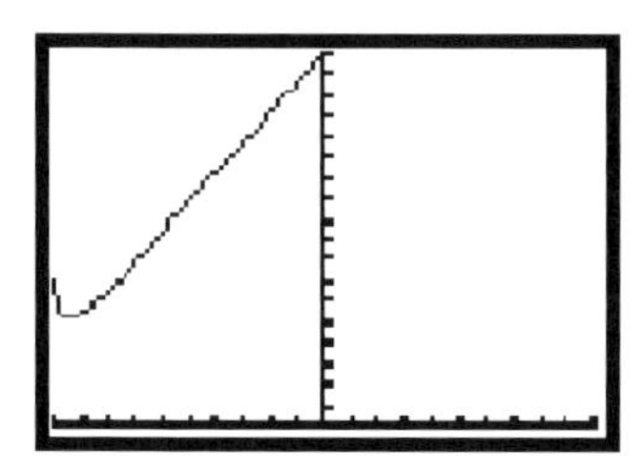

Plotting the relationship for $Y$ in the case of two internal reflections shows that the range of values of $Y$ is from about 52° to 180° ($Y$ is positive for negative values of $X$).

(b) The minimum value of $Y$ is 51.91° (to 2 d.p.).

(c) There is bunching in the values of $Y$ near 52° ($X$ about $^{-}0.95$). However, the bunching is not quite so pronounced as in the case of the primary bow.

(d) The angle $Y$ for the secondary bow will be just under 52°, for Descartes' value of $k$.

## Activity 18

The angles for the red and violet bands in the secondary bow are found by looking for the minimum values of the function

$$Y = 180° + 6\sin^{-1}(kX) - 2\sin^{-1} X$$

with the appropriate values of $k$. The angles at which you would expect to see the red and violet bands turn out to be about 51° and 54°, respectively.

## Activity 19

What you write for this activity will obviously be very individual, but here are some general observations.

(a) Many people will have had little idea about mathematical modelling at the start of MU120. You may well have had experience of the use of mathematics, but modelling frameworks may have been new. Now you have seen the mathematical modelling and statistical investigation cycles in use in the context of historical attempts to explain the rainbow phenomenon.

(b) In this section you have used algebraic and graphical representations of mathematical functions, including regression functions obtained from given data. You have employed geometry as well as trigonometry to find the functions that give the angles for the primary and secondary rainbows, and you have used the concepts of ratio and index in considering refraction.

How much you knew at the beginning of the course will depend upon your previous experience, but from the preparatory material, you may have been familiar with simple geometry (but may not have used it in three-dimensional contexts), word formulas (rather than algebraic formulas), numerical ratios and some basic ideas about a few mathematical functions corresponding to keys on your calculator (for example, for finding squares and square roots). However, before this course, you may not have made much use of trigonometric functions and their inverses, beyond in right-angled triangles, while the idea of the sine function representing a wave may be completely new to you.

(c) For most students, using a graphics calculator to plot data and draw graphs of mathematical functions will have been new—you may have previously done these things by hand. Storing data in lists and transforming the data using list arithmetic (for example, when converting from degrees to radians) may well have also been unfamiliar. Similarly, fitting the 'best' function to given data may have been new to you—you may have fitted a straight line to data points by eye, but not been aware of the full extent of possible regression functions and when they might be appropriate.

(d) The topics in Part 1 and aspects for Part 2 of TMA04 will change from year to year, but some of the activities in this section are likely to be useful.

## Activity 20

(a) The steps required in the program will be: input weightings; input scores; calculate assessment score; display this (to 2 s.f.).

(b) If your aim is to keep the program *short*, an ideal command to use is **Prompt**, which will contain all the weighting inputs and all your CMA/TMA score inputs within two single commands. Here is a suitable program, SCORE1, based on **Prompt**.

| PROGRAM: SCORE1<br>Fix 2 | Set the calculator output to 2 d.p. |
|---|---|
| Disp"GIVE WEIGHTINGS" | Display instruction. |
| Prompt A,B,C,D,E,F,G,H | Input weightings into A,B,C,D,E,F,G,H. |
| Disp "GIVE YOUR SCORES" | Display instruction. |
| Prompt I,J,K,L,M,N,O,P | Input scores into I,J,K,L,M,N,O,P. |
| (AI+BJ+CK+DL+EM+<br>FN + GO + HP)/100 → Q | Calculate score and store in Q. |
| Disp "YOUR SCORE IS", Q | Display score. |

The disadvantage of **Prompt** is that it cannot be used in association with explanatory text and so requires the user to remember what the letters A, B, C, ... and I, J, K, ... correspond to. Thus Input A corresponds to the first weighting (for CMA42), Input B to the weighting for CMA43, and so on. Correspondingly, Input I refers to your first assignment score (for CMA42), Input J to your assignment score for CMA43, and so on.

If having a *clearly explained* program is your priority, you might use the less space-efficient command, **Input**, as in SCORE2.

| PROGRAM: SCORE2<br>Fix 2 | Set the calculator output to 2 d.p. |
|---|---|
| Disp "GIVE WEIGHTINGS" | Display instruction. |
| Input "CMA42?",A<br>Input "CMA43?",B<br>Input "CMA44?",C<br>Input "CMA45?",D<br>Input "TMA01?",E<br>Input "TMA02?",F<br>Input "TMA03?",G<br>Input "TMA04?",H | Input weightings into A,B,C,D,E,F,G,H. |
| Disp "GIVE YOUR SCORES" | Display instruction. |
| Input "CMA42?",I<br>Input "CMA43?",J<br>Input "CMA44?",K<br>Input "CMA45?",L<br>Input "TMA01?",M<br>Input "TMA02?",N<br>Input "TMA03?",O<br>Input "TMA04?",P | Input scores into I,J,K,L,M,N,O,P. |
| (AI + BJ + CK + DL + EM +<br>FN + GO + HP)/100 → Q | Calculate score and store in Q. |
| Disp "YOUR SCORE IS", Q | Display score. |

(c) Here is a set of instructions:

Press the PRGM key and scroll down to the program name SCORE1 (or SCORE2).

Press ENTER to paste to the home screen.

Press ENTER to run the program.

The program asks first for the assignment weightings (as in the *Course Guide*). Input each as requested, and press ENTER.

The program then asks for your assignment scores. Input each one, and press ENTER.

The program will then calculate and display your assessment score.

(d) You might need to change the number and names of the assignments. However, you might be able to produce a more sophisticated program using two loops and lists which would cover any number of assignments.

## Activity 21

(a) The program should include the following basic steps:

Generate the two numbers for the problem—the number that is to be increased, and the number for the percentage increase. These could be generated using random numbers—decide on the range for each number and the command to be used to generate the number. Allocate variable names to these numbers (for example, A and B).

Display the problem on the calculator screen (consider which display command(s) to employ), using sufficient words to tell the user what to do: for example, increase A by B%.

Ask the user to input his/her answer (the instruction might say 'Input your answer correct to the nearest whole number'), and allocate a variable name, say C, to this answer.

Check the user's answer by doing the calculation and comparing the correct answer with the input from the user (the IF command might be helpful here).

Tell the user whether or not his/her answer is correct.

(b) The instructions should cover: how to find and run the program; and how to calculate a given percentage increase (explaining the technique). At each step, an idea of what to expect on the calculator screen should be given.

(c) Additional features might include giving the user some choice of the level of difficulty: for example, simple level (all numbers to be multiples of 10), medium difficulty (all numbers to be multiples of 5), more difficult (any whole numbers), and advanced (includes negative numbers, decimal numbers and decreases as well as increases). You might also give instructions at the end about how to repeat the program or quit.

## Activity 22

(a) You might like to add your programming skills to your notes on the activity sheet.

(b) Your answer will be individual to you. But when deciding which skills are relevant to the TMA04 Part 2 aspect, be very careful to check exactly what is required by the question and to ensure that each point you make is relevant to the given aspect.

## Activities 23–25

Everybody's responses will be different.

## Activity 26

You may not have produced a list as long as the one below, and you may have included some topics which are omitted from the list. However, the list may remind you of topics you had forgotten.

(a) *Unit 3*
Earnings ratios and indices; median; quartiles; interquartile range and range of batches of data.

*Unit 4*
Standard deviation, randomness of data; cause and effect in numerical relationships.

(b) *Unit 6*
Relationships between distances, angles and areas on the ground, and the representation of these on a map; map scales or ratios; relationships between grid bearings and magnetic bearings.

*Unit 7*
Conversions between measurements in different units; direct proportionality relationships; the connections between gradient, rates of change, and speed.

*Unit 8*
The distributative and associative properties of numbers. Formulas and tables for conversions between different units.

*Unit 9*
Relationship between the string-length ratio and the musical interval between the notes produced.

*Unit 12*
Simple interest; compound interest; APR; the relationship between interest rates, repayments and the amount owing on a loan.

*Unit 13*
Proportionality relationships; scaling a recipe; relationship between the scaling of lengths and the areas and volumes dependent on them.

*Unit 14*
Relationships between the lengths of the sides of similar triangles; trigonometric ratios.

*Unit 16*
Relationship between angle of incidence and angle of refraction; refractive index.

## Activities 27–33

There are no comments on these activities as the topics will change from year to year and your answers depend on your choice of questions and aspect.

## Acknowledgements

### Cover

Indian rock art rainbows: Kenneth Sassen and Optical Society of America.
Other photographs: Mike Levers, Photographic Unit, The Open University.

# Index